1990

THE GEOMETRY
OF GENETICS

WILEY-INTERSCIENCE MONOGRAPHS IN CHEMICAL PHYSICS

Editors

Sean P. McGlynn, *Louisiana State University*
Gary L. Findley, *Louisiana State University*

Group Theory in Spectroscopy—With Applications to Magnetic Circular Dichroism: Susan B. Piepho and Paul N. Schatz

The Geometry of Genetics: Ann M. Findley, S. P. McGlynn, and Gary L. Findley

THE GEOMETRY
OF GENETICS

A. M. Findley, S. P. McGlynn, and G. L. Findley

Louisiana State University
Baton Rouge, Louisiana

WILEY

A Wiley-Interscience Publication

JOHN WILEY & SONS

New York • Chichester • Brisbane • Toronto • Singapore

Library of Congress Cataloging in Publication Data:

Findley, A. M. (Ann M.)
  The geometry of genetics/A. M. Findley, S. P. McGlynn,
  and G. L. Findley.

  p. cm.—(Wiley-Interscience monographs in chemical physics)
  "A Wiley-Interscience publication."
  Includes bibliographies and index.
ISBN 0-471-05617-0

1. Molecular genetics—Mathematics. 2. Molecular genetics.
3. Chemical evolution—Mathematics. 4. Chemical evolution.
5. Geometry. I. McGlynn, S. P., 1931–   . II. Findley, G. L.,
1952–   . III. Title. IV. Series.
QH438.4.M33F56 1989
574.87′328′01516—dc19   87-33982
                                  CIP

Printed in the United States of America

10 9 8 7 6 5 4 3 2 1

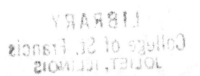

Preface

We began the research that comprises this monograph in 1977. During the period of 1978–1985 we published some ten papers on genetic coding theory and on the application of differential geometry to molecular genetics, in venues ranging from JTB to PNAS; one of us (GLF) wrote his PhD dissertation on this subject; and we presented approximately thirty invited lectures and contributed papers on theoretical biology at various universities and conferences in United States and abroad. We point this out only so that we can immediately note the following: During this same period we were also actively engaged in the study of atomic and molecular physics and, in particular, in molecular spectroscopy research.

Our secret, then, is out: Although this book is about biology, the three of us are really chemical physicists who, incidentally, have had—and continue to have—a deep interest in molecular genetics. (We also presume to hope that we have had some interesting *ideas* in molecular genetics.) The origin of this interest is not hard to discern. Each of us was originally trained in molecular biology/biochemistry at the graduate level, and one of us (AMF) even went so far as to earn her PhD in experimental molecular physiology. That we all eventually migrated to chemical physics is probably a result of our perception that biology is, in a fundamental sense, too difficult; thus we sought an area of research that would allow us the luxury of precision, as well as the freedom to attempt to inject some of this precision into biology. This book constitutes a summary of our efforts in this direction, and a review of some of the efforts made by others.

This monograph is inherently interdisciplinary: The subject is molecular genetics and evolutionary biology; the machinery that we develop relies heavily on mathematics; and the motivation for our approach is rooted in physics, at least insofar as our emphasis is on a search for "hidden symmetries" in biology. In order to make this work accessible to two different audiences (biologists and physical scientists), we have adopted the following tactic. After a general introduction, we divide our subject into three parts: structure, statics, and dynamics. Each of these parts is further subdivided into (1) a presentation of the relevant mathematics, (2) a description of the biological problem, and (3) a mathematical reformulation of the biological problem. We have set out to provide, in effect, basic mathematical and biological primers for each of the new developments that we discuss. (This is not to say that those readers who are unfamiliar with either the mathematics or the biology or, perhaps, both, will find the going particularly easy; we have, however, made a strong effort to smooth the way.)

Structure concerns the genetic code and, more specifically, an extension of genetic coding theory that we have termed the "generalized genetic code." *Statics* represents a realization of the basic processes of molecular genetics—replication, transcription, and translation—as operators on a certain linear space. And, finally, *dynamics* provides a differential geometric treatment of biological evolution, phrased at the molecular level. This hierarchy of complexity is intentional. Thus, statics is built on structure, and dynamics is built on both structure and statics. One advantage of such an organization of the subject is the exposure of an explicit, nested cohesiveness.

We hope that this book will be of use to biologists who have an interest in the construction of fundamental theories of biological phenomena. While we make no claims of "ultimate truth" for any of the theoretical ideas presented here, we do feel that our approach to theory construction in biology is generally valid and paradigmatic. We also hope that this work will be of interest to the ever-increasing number of chemists, physicists, and applied mathematicians who are becoming involved in biological problems.

Our research in molecular genetics has received continuous financial support from the U.S. Department of Energy, for which we are very grateful. We especially wish to thank Dr. Frank P. Hudson (formerly of DOE) for his interest and encouragement during the early stages of this work. And we are grateful to Dr. Matesh Varma and Dr. Robert Wood for continuing that interest and encouragement.

The manuscript for this book was originally solicited for the Wiley-Interscience Chemical Physics Series by the late Professor John B. Birkes of the University of Manchester. We hope that he would have been pleased with the final result.

This monograph was begun and completed at Louisiana State University; two of us (AMF and GLF), however, spent an interregnum on the chemistry faculty of New York University. We are grateful to Professor Edward NcNelis, Chairman of the NYU Chemistry Department at that time, for the many kindnesses that he showed to us. Also, parts of this book were written while AMF and GLF were in residence at the Hamburger Synchrotronstrahlungslabor (HASYLAB) (DESY, Hamburg, FRG). We thank Dr. Volker Saile (Vice-Director of HASYLAB) for his warm hospitality.

Dr. Theodore P. Hoffman of Wiley-Interscience was most patient during the long gestation and difficult labor of this project. We appreciate his courtesy.

Finally, we thank our wives, husbands, children, parents, and so on: the distractions which they provided were, in the end, contributions.

A. M. FINDLEY
S. P. MCGLYNN
G. L. FINDLEY

Baton Rouge, Louisiana
March 1988

Contents

PART 4.   DYNAMICS

THE GEOMETRY
OF GENETICS

PART 1   INTRODUCTION

1 General Overview

1. Prefatory Comments

Genetics is that area of biology which investigates heredity and variation. At the molecular level, the hereditary information content of a gene is vested in the linear arrangement of the nucleic acid bases in a deoxyribonucleic acid (DNA) molecule. This essential tenet, first proposed by Watson and Crick [1], has provided the genesis for numerous insightful studies into the structure–function relationship of the hereditary process. Indeed, the ensuing 30 years has witnessed a veritable explosion of experimental investigation, both chemical and biological in nature, which has enhanced our knowledge of the fundamental processes that comprise what has come to be known as *molecular genetics*.

As a discipline, molecular genetics encompasses a diverse assembly of subject matter, which ranges from the purely chemical to the purely biological. In every case, however, an underlying thesis is discernible, namely, that the structural and functional characteristics of an organism are directly determined by molecular interactions at the subcellular level. Thus molecular genetics represents a redirection of biological reasoning from the level of the organism to the level of the complex physicochemical processes which occur within all organisms. At the cornerstone of this enterprise is the so-called central dogma, that is, that DNA *replication*, DNA *transcription* into messenger ribonucleic acid (mRNA), and mRNA *translation* into protein serve as the molecular basis for macroscopically observable heredity. A significant amount of experimental effort has been expended toward the explication of the enzymic reactions which constitute these processes. What has emerged is an overall picture which, although it remains incompletely understood, appears to be remarkably constant throughout all living systems. It is to this emergent constancy that we offer a systematic explanation.

2. Description of Intent

In the following exposition we consider the discipline of molecular genetics to consist solely of a compilation of *conceptual* processes which comprise the

3

biological-information storage, retrieval, and processing systems. We eschew any discussion of the intricate molecular interactions which underlie these biological-information systems, except to the extent necessary to delineate the specific phenomena involved. Consequently, attention is concentrated only on those aspects of molecular genetics which are amenable to abstract formulation without *explicit* recourse to physical consideration. As a result, the following presentation is rigorously external to the purview of traditional biophysics, since we fixate on the conceptual foundations of molecular genetics rather than on the underlying physics.

Our intention in this work is to present a mathematical theory of molecular genetics predicated solely on biological concepts. Principal motivation for such an effort is the expression of molecular genetics in a formalism that is totally amenable to the construction of significant and meaningful questions concerning the relevance of symmetry considerations in biological systems.

Of course, the notion of biological symmetry is unavoidably fuzzy. Nevertheless, questions pertaining to biological symmetry are critically important because the detection of such inherent symmetries is comparable to the discernment of biological laws. With the introduction of a mathematical formalism, however, the question of symmetry is stripped of its imprecision. Thus a realization of a part of biology as mathematics is, in essence, the beginning of research that may ultimately lead to the explication of biological laws. Such laws may be every bit as well founded empirically, and as logically consistent, as physical laws. In addition, in biology the potential exists to construct a symmetry-searching apparatus in advance of a genuine demonstration of the existence of such symmetries. The rewards of such a program rest in the possibility of guiding research, through posing questions which arise in a natural fashion from the mathematics, in the perception of biological laws.

3. Chapter Summaries

The results which we present constitute a mathematical formulation of two aspects of molecular genetics. First, we investigate the symmetry of the genetic code from a set- and group-theoretic viewpoint. Second, we explore the realization of molecular evolution in a Riemannian geometry. In communicating this work we appreciate that we are addressing two potentially disparate audiences: those familiar with the molecular biology but largely unaware of the requisite mathematics and, conversely, those fully competent in the mathematics but uninitiated with respect to the biology. Therefore, with clarity of exposition in mind, the body of this work is apportioned into three self-contained parts, appropriately entitled Structure, Statics, and Dynamics. Each of these is dissected further into three chapters. In each case, the first chapter briefly explores the relevant mathematical foundations basic to our subsequent development. Axiomatic systems are introduced primarily without obvious motivation, and no proofs are given. A second chapter provides a précis of some fundamental concepts from molecular genetics. Again, we do not claim any completeness of exposition. Indeed, we discuss only those subjects for which we subsequently furnish a mathematical interpretation. Finally, we conclude each part with a "synthetic" chapter which details a mathematical formalism designed to explicate the relevant biological material.

Preceding these three parts, we have chosen to include an introductory section (Part 1) which provides both a philosophical (Chapter 1) and a biological (Chapter 2) overview of our work. Chapter 2 is intended as a general account of the current theory of molecular genetics. We offer only a sketch of the conceptual foundations of this theory and refer the interested reader to standard works and subsequent chapters (i.e., Chapters 4, 7, and 10) for detailed empirical justifications of these concepts. However, one should bear in mind that Chapter 2 represents our choice of what we feel to be centrally important to the theory of molecular genetics, and this, in turn, forms the basis for our later mathematical abstractions.

Part 2, entitled Structure, is divided into three portions. Chapter 3 furnishes a detailed introduction to the theory of modern algebra, with particular emphasis placed on those set- and group-based constructions pertinent to our subsequent investigations. This material also fixes certain notational conventions which we have adopted. Chapter 4 supplies a narrative of the current biological wisdom concerning the genetic code. Finally, as a synthesis, Chapter 5 explores the mathematical formulation of the symmetry properties of the genetic code. Briefly, necessary (but not sufficient) conditions for codon degeneracy are found from a consideration of the so-called ambiguous or alternative codon assignments. The argument, elaborated by a group-theoretic construction, shows that these ambiguous codon assignments are perfectly valid from symmetry considerations and need not be considered as mistakes of the biological system. The result is an expanded view of the nature of the genetic code, which is at significant variance with that embodied in the "central dogma" of molecular genetics.

Part 3, entitled Statics, is concerned with an initial geometric description of biological information transfer. In Chapter 6 the previous discussion of abstract algebra (Chapter 3) is extended to provide the necessary information for the formation of vector spaces and metric spaces. Chapter 7 details the intricate molecular processes of transcription (the production of single-stranded mRNA from a double-stranded DNA template) and translation (the synthesis of a protein from an mRNA template). These concepts are brought together in Chapter 8, with the result being a realization of molecular genetics as an n-dimensional vector space. DNA, RNA, and protein molecules are treated as vectors in separate finite-dimensional spaces. Transcription then becomes a linear operator mapping DNA space onto RNA space, while translation is treated as a linear operator mapping a subset of RNA space onto protein space. Two separate vector space formulations, different only in field structure, are investigated: (1) an n-dimensional Euclidean vector space over the real field, taking as a basis the information units of the molecules in question (DNA codons, RNA codons, or amino acids) and (2) an n-dimensional vector space over a finite field, where the elements of the field are chosen (by order isomorphisms) to be the DNA (or RNA) bases. By arguments elaborated in Chapter 7, one is led to choose formulation (1) as the more appropriate structure within which to phrase questions pertaining to the *dynamic* (evolutionary) nature of molecular genetics. Addition of an evolutionary time coordinate to the space of DNA vectors gives rise to the *informational space–time manifold*. This manifold, however, suffers from an inherent information loss which is derivative of the nature of the informational space of DNA vectors, namely, the vectors that represent DNA molecules differing only in codon order lie at zero distance from one another and cannot be distinguished. In attempting to remove this artifact in a nontrivial manner, one is lead to relax the Euclidean constraint on the informational space–time manifold.

This relaxation is accomplished in Part 4, entitled Dynamics. In Chapter 9 a development of local differential geometry is presented along classical lines. Again, Riemannian geometry is defined and investigated from a purely mathematical viewpoint, and certain notational conventions are fixed. The biological postulate that evolutionary motions on the informational space–time manifold are geodesics is made in the final section of this chapter. The result of this postulate is the interpretation of the geodesic equations as *evolutionary equations of motion*. These equations of motion are defined with respect to an *evolutionary field* which arises from the intrinsic structure of the informational space–time manifold. (The analogy to concepts from general relativity is explicitly stated.) Thus when realized as a Riemannian geometry, molecular genetics may be interpreted as a biological field theory. This field theory may be made specific by a choice of the intrinsic structure of the informational space–time manifold, and such a choice defines a specific *genetic cosmology*. The fundamental result is that the solution to those evolutionary questions which are formulated at the DNA (or RNA, or protein) level resides, in principle, in the knowledge of the biologically correct genetic cosmology.

To provide further biological insight into the problem we are attempting to treat mathematically, Chapter 10 furnishes a survey of the relevant, and often divergent, concepts concerning macromolecular evolution. Finally, the very general development of Chapter 9 is made more concrete in Chapter 11 by an investigation into methods of research in genetic cosmology. An (incomplete) model genetic cosmology, in which the intrinsic field is a function of only the evolutionary time, is presented and the evolutionary equations of motion for a weak evolutionary field, which is a function of only the evolutionary time, are derived. In addition, it is shown how the evolutionary information loss, inherent in the Euclidean vector space formulation of Chapter 8, may be alleviated in a curved informational space–time manifold. In conclusion, the nature of future empirical input into genetic cosmology is discussed.

Reference

1. J. D. Watson and F. H. C. Crick, *Nature*, **171**, 964 (1953).

2   Biological Overview

Genetics is that area of biology which investigates heredity and variation. At the molecular level it is concerned with the complex physicochemical basis for information storage, retrieval, and processing systems. In this introductory chapter we provide a synopsis of some fundamental insights that molecular genetics has contributed to our understanding of the hereditary process. Considering the spectacular advances which have occurred in this field, and the concomitant literature explosion, our aims must be modest. Therefore, we do not claim any completeness of exposition. Rather, we shall only present brief sketches of the structures and transformations of the appropriate informational macromolecules. More systematic treatments, illustrated by specific biological systems, will be found in standard texts [1–6] and, in part, in subsequent chapters (Chapters 4, 7, and 10). In any case, the reader should bear in mind that this overview reflects our choice of what we feel to be centrally important to that portion of the biological theory of molecular genetics to which we shall later make mathematical application.

A further caveat is also appropriate at this point. Much excitement has arisen recently concerning the differences between prokaryotic and eukaryotic molecular organization of their respective genetic systems [4, 7]. Undoubtedly, future research will continue to require us to complicate our simplistic prokaryotic picture of molecular genetics to make it fully amenable to eukaryotic organisms. Our view is that the prokaryotic theory stands as a zeroth-order approximation to the eukaryotic theory, and, on the strength of this position, our sketch of molecular genetics will often eschew any reference to the eukaryotic, first-order corrections. With the above limitations in mind, we turn first to a discussion of the structure of DNA.

1.   DNA Structure

A gene is a length of nucleic acid which is responsible for the transmission and expression of a hereditary characteristic. DNA serves as the repository of information which determines the genetic variability of an organism. This information is encoded in the arrangement of nucleic acid bases (T ≡ thymine,

A ≡ adenine, C ≡ cytosine, G ≡ guanine) along the polymer chain. There is a linear relationship between the nucleic acid base sequence of a gene and the amino acid sequence of the polypeptide chain synthesized through the mediation of that gene. To comprehend fully the nature of the correspondence between a gene and its resultant polypeptide chain, we must first consider the molecular structure of DNA.

DNA is a polymeric molecule composed of nucleotide subunits. A nucleotide may in turn be dissected into its constituent members: a five-carbon 2-deoxyribose sugar moiety, phosphoric acid, and one of four nitrogen-containing heterocyclic bases—adenine and guanine (purines) or cytosine and thymine (pyrimidines) (see Figure 2-1). In 1953, Watson and Crick deduced the geometric arrangement of these components along a stretch of DNA [8]. Using X-ray diffraction data of DNA fibers provided by Franklin and Wilkins, they constructed the model structure shown in Figure 2-2. DNA is composed of two right-handed intertwined helical polynucleotide chains (a double helix or duplex) coiled around the central or long axis of the molecule.

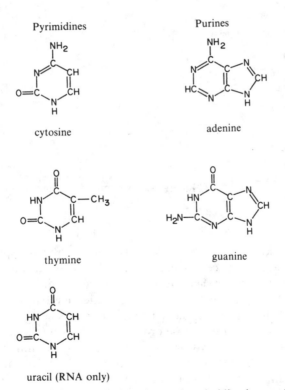

Figure 2-1. Structure of the principal purine and pyrimidine bases of nucleic acids.

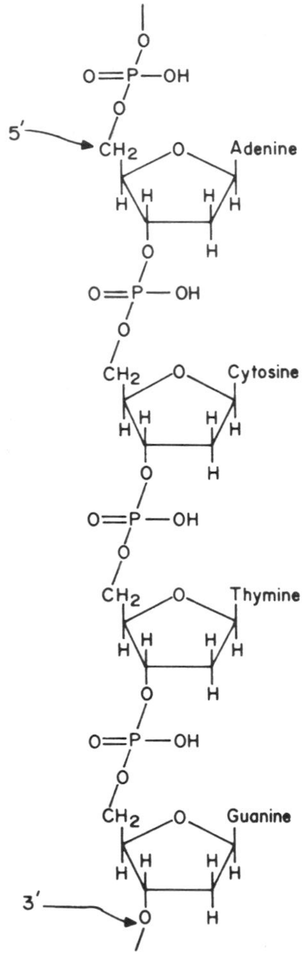

Figure 2-2. A segment of single-stranded DNA with 5′, 3′ positions indicated.

The DNA backbone consists of alternating sugar and phosphate moieties (arranged parallel to the central axis of the helix) with phosphate adjoined to the 5′-carbon of the sugar via a covalent 3′, 5′-phosphodiester bond. From the 1′-carbon of the pentose substituents, the nucleotide bases extend into the center of the helix (i.e., perpendicular to the long axis of the molecule). Bases from opposite strands are in close proximity and form strictly specified base pairs stabilized through hydrogen bonds between the amino or hydroxyl hydrogen and the keto oxygen or imino nitrogen of purine and pyrimidine bases. In addition, hydrophobic interactions contribute to the stability of DNA structure by maintaining the orientation of the nonpolar bases to the interior of the helix, away from the aqueous cellular environment, while favoring the placement of the polar sugar–phosphate backbone to the outside, in contact with water.

Chargaff's equivalency rule [9] established a stoichiometric relationship among bases in a native DNA molecule: [G] = [C] and [A] = [T] ([   ] read: concentration of). Consistent with this observation, Watson and Crick [10] postulated that base pairing occurs for guanine with cytosine (via three hydrogen bonds) and for adenine with thymine (via two hydrogen bonds). The pairing of purine with pyrimidine bases ensures that the molecule will possess a uniform diameter of approximately 20 Å.

As a consequence of this model, several unique features of the DNA molecule may be deduced. The base sequence of the two strands of DNA are necessarily complementary (i.e., the sequence of one strand may be converted to that of the other by performing the appropriate substitutions: A → T, G → C, T → A, C → G). The DNA molecule, then, is self-complementary, carrying two complete sets of information, albeit in complementary notation (i.e., they possess opposite sense).

In addition, a single strand of DNA possesses a built-in chemical polarity (5', 3') attributable to the substituent positions on the deoxyribose unit of the DNA polymer (see Figure 2-2). The 5' position refers to the hydroxy-substituted carbon external to the deoxyribose ring, while the 3' position refers to the hydroxy-substituted carbon internal to the ring. In a single strand of DNA, adjacent deoxyribose units are associated via a phosphodiester linkage, with the bonding occurring at the 5' position on one unit and at the 3' position on the other. If we now consider an entire single-stranded DNA, one end of the molecule terminates in a deoxyribose unit which is bonded to the rest of the molecule at the 3' position, while the other end terminates in a unit bonded to the rest of the molecule at the 5' position. These are designated the 5' and 3' ends of the molecule, respectively. In a double-stranded DNA, however, association occurs such that the two strands of DNA have opposite polarity (i.e., they are antiparallel) if we view the molecule, left to right say, along the length of the chain. In other words, at the left end of the polymer one strand has a 3' terminus while the other has a 5' terminus; at the right end, the former strand has a 5' terminus while the latter has a 3' terminus. These distinctions are critically important for transcription and translation (see Chapter 2, Sections 3 and 4, and Chapter 7).

2. DNA Replication

The structure of the DNA molecule also led Watson and Crick to postulate a mechanism by which it could be replicated accurately [10]. This, in turn, developed into what is commonly termed the central dogma of molecular genetics. According to the central dogma, there are three fundamental phases in genetic information processing:

1. *Replication* (DNA serves as a template for additional DNA synthesis)
2. *Transcription* (DNA provides a template for mRNA production)
3. *Translation* (mRNA furnishes a template for protein synthesis)

That is,

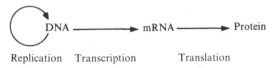

DNA biosynthesis occurs during cellular division or reproduction. To ensure the precise replication of DNA, or any other informational macro-molecule, DNA must function as a template for the synthesis of two replicate molecules which are functionally identical to (i.e., possess the same base composition as) the parent DNA. Replication is a complex enzymic reaction which is, as of yet, incompletely understood from a chemical standpoint. In conceptual terms, however, the overall process may be described as follows.

Given the complementary nature of a double-stranded DNA molecule, Watson and Crick proposed [10] that replication proceeds first through the unwinding of the strands to expose two single DNA chains. Each of these chains then serves as a blueprint, dictating the nucleotide base sequence for the synthesis of a new complementary strand, with synthesis directed from the 3' to the 5' ends of the single-stranded template. These two new complementary strands, each associated with one of the parental chains, form the two replicate DNA molecules. This type of synthesis is termed *semiconservative*, since the parent DNA is entirely contained in the product DNAs: one of the parent strands is found in one replicate molecule, while the other parent is located in the second replicate (see Chapter 7, Section 1, and [11]).

The fidelity of DNA replication is reinforced by the structural stability inherent in the complementarity of base pair formation. However, it is also true that DNA synthesis is not without some potential for error. Indeed, it is at the level of replication that genetic mutation is presumed to operate. The occurrence of such mutations is simply a reflection of the reality that DNA replication is not entirely faithful. In principle, three basic types of mistakes may arise.

1. *Substitution.* This occurs from a mismatch in base pairing during the formation of the new complementary strands and results in the

substitution of one base pair for another at a particular point in the molecule.

2. *Deletion.* This is the loss of a specific base pair from a particular point in the molecule.

3. *Insertion.* This is the addition of a specific base pair at a particular point in the molecule.

Deletions and insertions are collectively known as phase-shift (or frame-shift) mutations at the level of translation (see below). Such mutations of the genetic material may have rather profound consequences for the total evolutionary progress of an organism. We will return to this subject in subsequent parts of this work.

3. DNA Transcription

The message encoded in the nucleotide base sequence of a DNA molecule eventually results in the formation of a specific protein. In fact, that segment of DNA which contains the information necessary to completely specify one protein (or, more precisely, a single polypeptide) is the molecular equivalent of a gene (see Chapter 2, Section 5). DNA does not act directly in protein synthesis, however. Instead, mRNA (a specific type of ribonucleic acid) is first synthesized from DNA, and subsequently serves as the actual template through which protein synthesis is mediated. The molecular structure of RNA differs from that of DNA in the following ways.

The sugar moeity of RNA is ribose. Ribose differs from deoxyribose in being hydroxylated at the 2′ position on the ring.

The DNA base T ≡ thymine is replaced by U ≡ uracil.

RNA is (usually) single stranded, whereas DNA is (usually) double stranded.

Transcription is the synthesis of an mRNA from a DNA template (see Figure 2-3). This process serves merely to transfer the genetic information contained in DNA to the mRNA molecule. There are two essential points to note concerning transcription:

1. Only *one* of the two complementary DNA strands may serve as the template for a specific mRNA.

2. Synthesis proceeds from the 3′ end to the 5′ end of the DNA template.

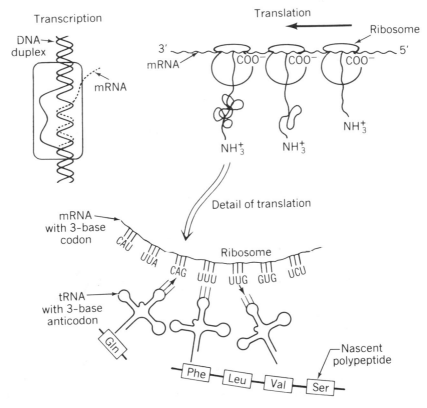

Figure 2-3. Schematic diagram of transcription and translation.

Although replication and transcription differ in the specifics of their respective chemical mechanisms, they are conceptually quite similar. During transcription, the double-stranded DNA unwinds, exposing the single DNA chains so that one may serve as the template. Synthesis then proceeds from the 3' end to the 5' end of the selected template, and during synthesis, the RNA base U pairs with the DNA base A. Of course, the mRNA molecule *grows* from the 5' to the 3' end. Upon completion, the newly formed mRNA is released from the DNA template, which then resumes its right-handed helical form while the mRNA is transported to the cytoplasm (of eukaryotic cells), the site of translation.

4. mRNA Translation

The biosynthesis of protein from an mRNA template is termed *translation*. During protein synthesis, the mRNA nucleotide base sequence is translated into the amino acid sequence of a nascent polypeptide via a 4-letter, 64-word *genetic code* (see Chapter 4). Each triplet (or *codon*) of RNA bases prescribes the introduction of one amino acid into a growing polypeptide chain. Thus the linear arrangement of codons along the mRNA template (which ultimately derives from the base sequence of the controlling DNA molecule) determines the linear arrangement and types of amino acids in the protein product.

A striking feature of the genetic code is that while there are $4^3 = 64$ possible codons, there are only 20 amino acids. Thus more than one codon specifies the same amino acid (i.e., the genetic code is *degenerate*). The symmetry characteristics of the genetic code, as exhibited by degeneracy, are of intrinsic interest, and this subject is developed extensively in Part 2 of this work.

Protein synthesis proceeds in the $5' \rightarrow 3'$ direction along the mRNA template (see Figure 2-3). Therefore the sense of a codon must be interpreted from the $5'$ to the $3'$ end of the mRNA molecule. The actual reading of the message template occurs within a ribosomal complex which serves, among other things, to set the phase of the message. At the ribosomal level, each codon is recognized by a particular transfer RNA (tRNA), which transports the amino acid specified by a certain codon to the site of protein synthesis. This recognition process involves base pairing between the mRNA codon and the tRNA anticodon. For example, if UAG is the codon in question, then CUA is the tRNA anticodon.

Synthesis is initiated at the AUG codon which, when acting as the initiator, codes for the introduction of N-formylmethionine. The ribosome then shifts one triplet down the mRNA molecule in the $3'$ direction, and the amino acid which is coded for by the new codon is brought into position by the appropriate tRNA. Next a peptide bond is formed between N-formylmethionine and the new amino acid. The ribosome then shifts down to a third triplet. This process is continued until one of the three termination codons (UAA, UAG, or UGA) is reached and protein synthesis is interrupted. The final product of translation is a specific protein having an N-formylmethionine residue at the amino end of the protein. Incidentally, this first residue is usually removed later during modifications which transform a newly synthesized protein into a "mature" one.

The initiator codon–ribosome complex serves to partition the mRNA base sequence into codons, that is, this complex formation determines the phase (reading frame) of translation. Phase-shift mutations (DNA deletions and insertions) act by modifying the reading frame. Consider the partial mRNA sequence

$$5'\cdots\text{-(AAA)(CCU)(CGC)A-}\cdots 3'$$

which codes for (see Chapter 4)

$$\cdots\text{-Lys-Pro-Arg-}\cdots$$

Deletion of the third base (from the 5' end) results in the sequence

$$5'\cdots\text{-(AAC)(CUC)(GCA)-}\cdots 3'$$

which now codes for

$$\cdots\text{-Asn-Leu-Ala-}\cdots$$

Thus deletion mutations can radically change the protein structure by altering the reading frame. (Insertion mutations act in a similar manner.) If equal numbers of deletions and insertions occur, however, then the phase remains unaltered posterior (in the 3' direction) to the last deletion or insertion.

5. Gene Structure and Function

Traditionally a gene is interpreted as that portion of the genetic material (i.e., the chromosome) which determines a particular hereditary characteristic. From a more molecular viewpoint, a gene has alternately been described as that segment of genetic material which codes for one enzyme [12], or, more generally, one protein (since not all proteins are enzymes), or, most generally, one polypeptide chain (since many proteins contain multiple polypeptide chains which, when dissimilar in amino acid sequence, are necessarily specified by more than one gene).

In any case, not every gene is expressed in terms of polypeptide chains. Some genes code for different RNAs (e.g., tRNA, ribosomal RNA), while others produce no detectable product whatsoever. Genes which dictate the synthesis of some final product, such as an RNA or a polypeptide, are termed structural genes. Genes which are not expressed through a gene product serve as regulatory genes. Such areas provide information concerning the location of the beginning or end of structural genes and aid in the on/off control of structural gene transcription.

With the advent of rapid DNA sequencing techniques [13, 14], an increasing number of DNA base sequences have been determined. Such experiments have indicated that certain DNA segments within all eukaryotic genes are coding fragments, called *exons*, while other DNA regions in genes are nontranslated, termed *intervening sequences* or *introns* [15–18]. The presence of introns within a gene disturbs the colinear correspondence between the base sequence of that gene and the ultimate amino acid sequence of its resultant polypeptide. While a definitive explanation remains lacking, introns have alternately been described as possible regulatory signals or as a means of partitioning large genetic messages into smaller, exchangeable subunits which may undergo genetic recombination during the evolution of species.

Also, unlike prokaryotes, which usually possess only one copy of DNA in each cell, segments of eukaryotic DNA exist in multiple copies. Satellite DNA is the term used to describe highly repetitive regions which are believed to be nontranslated segments. Moderately repetitive sequences are thought to be regulatory in nature, while unique single-copy regions (or regions with very few copies) appear to be structural genes [19].

6. Mitochondrial DNA

In addition to their nuclear DNA, eukaryotic cells also possess small amounts of cytoplasmic DNA which is localized in the mitochondria, or chloroplasts of plant cells. Mitochondrial DNA (mtDNA) exists as a relatively small, circular duplex, which may represent the remnants of the genome of primitive bacteria that entered the cytoplasm of host cells and established a symbiotic existence, eventually evolving into the modern-day organelle [7]. At present, mtDNA codes for mitochondrial tRNAs, ribosomal RNAs (rRNAs), and a limited number of mitochondrial proteins. However, since most mitochondrial proteins ($\sim 95\%$) are specified by nuclear DNA, the underlying reason for the existence of mtDNA remains an interesting subject for speculation.

7. Summary

The molecular genetic system serves a twofold purpose:

1. It transmits information from one generation to the next (DNA replication).
2. It expresses this information within one generation (transcription and translation).

Each of these processes is composed of complex enzymic reactions which are neither well understood nor even, in many cases, well characterized. The basic conceptual nature of information storage and retrieval, however, appears to be quite secure. It is this conceptual system with which we shall be concerned in later chapters.

References

1.   J. Cairns, G. S. Stent, and J. D. Watson, Eds., *Phage and the Origins of Molecular Biology*, Cold Spring Harbor Laboratory, Cold Spring Harbor, NY, 1966.
2.   W. Hayes, *The Genetics of Bacteria and Their Viruses*, 2nd ed., Wiley, New York, 1968.
3.   J. D. Watson, *Molecular Biology of the Gene*, 3rd ed., Benjamin, New York, 1976.
4.   G. S. Stent and R. Calendar, *Molecular Genetics, an Introductory Narrative*, 2nd ed., Freeman, San Francisco, CA, 1978.
5.   F. Ayala and J. Kiger, *Modern Genetics*, Benjamin–Cummings, Menlo Park, CA, 1980.
6.   L. L. Mays, *Genetics: A Molecular Approach*, Macmillan, New York, 1981.
7.   See, for example, J. F. Fredrick, Ed., *Origins and Evolution of Eukaryotic Intracellular Organelles*, in *Ann. N. Y. Acad. Sci.*, **361** (1981).
8.   J. D. Watson and F. H. C. Crick, *Nature*, **171**, 737 (1953).
9.   E. Chargaff, *Experimentia*, **6**, 201 (1950).
10.   J. D. Watson and F. H. C. Crick, *Nature*, **171**, 964 (1953).
11.   M. Meselson and F. W. Stahl, *Proc. Natl. Acad. Sci. USA*, **44**, 671 (1958).
12.   G. W. Beadle and E. L. Tatum, *Proc. Natl. Acad. Sci. USA*, **27**, 499 (1941).
13.   F. Sanger, S. Nicklen, and A. R. Coulson, *Proc. Natl. Acad. Sci. USA*, **74**, 5463 (1977).
14.   F. Sanger, *Biosci. Repts.*, **1**, 3 (1981).
15.   F. H. C. Crick, *Science*, **204**, 264 (1979).
16.   J. Abelson, *Ann. Rev. Biochem.*, **48**, 1035 (1979).
17.   C. C. F. Blake, *Nature*, **291**, 616 (1981).
18.   P. Borst and L. A. Grivell, *Nature*, **289**, 439 (1981).
19.   E. H. Davidson and R. J. Britten, *Science*, **204**, 1052 (1979).

PART 2 STRUCTURE

3   Sets and Their Structures

One of the most fruitful generalizations ever made in mathematics, the concept of *set*, was first introduced and exploited by Georg Cantor in the late nineteenth century. Today it is hard to overemphasize the pervasive nature of set theory in modern mathematics. Armed with sets, and the idea of *structure* on a set, it becomes possible to relate the various special constructions of mathematics (e.g., group, ring, field) to one another. This lends a sweeping unity to the study of sets and their structures, a unity that permits one to start from a relatively uncomplicated construction (a group, say) and to proceed to a relatively complicated construction (a field, say) simply by imposing a more detailed structure on the underlying set.

In a certain sense, a change of structure on a set is nothing more than a change of view as to what the base set is supposed to "mean." Alternatively (and, really, more correctly), the structure *is* the meaning. This implies, then, that when one applies mathematics to a physical system, the suitability of the application is contingent upon the aptness of the structure chosen. Such, indeed, is the overall theme of this monograph.

Our intent in this chapter is to introduce sets, and to define and discuss their basic structures. We shall find later (Chapter 5) that these fundamental mathematical constructions provide a natural language for the discussion of a fundamental biological construction, namely, the genetic code.

1.   Sets

Fundamental to all of our impending discussions is the concept of a set. Within the "naive" approach to set theory that we shall adopt, sets are not defined—they are intuitively grasped. Thus by the term *set* we shall understand a collection of distinct objects gathered together by some common characteristic to form a conceptual *whole*. These distinct objects comprise the elements of a set. The essential property of a set is *inclusion*; that is, whether or not a given element is a member of a specific set.

Let A be a set and x an element which may or may not be contained in A. Notationally, either $x \in A$ (read: x is an element of A) or alternatively $x \notin A$, also written $x \in' A$ (read: x is not an element of A). The elements of a set are generally enclosed within braces { } to indicate set-theoretic inclusion.

Obviously it might prove rather tedious or even impossible to list all of the members of a set. In this case, a shorthand representation is useful. For example, the set of natural numbers $\mathbb{N}$ may be extensively presented as the series $\mathbb{N} = \{1, 2, 3, \ldots\}$. Alternatively, we can construct an intensive definition of $\mathbb{N}$ as follows: $\mathbb{N} = \{x: \text{if } x \in \mathbb{N}, \text{ then } f(x) \in \mathbb{N}\}$, where $f(x)$ means "the integer successor of x." Now if we let the quantity to the right of the colon (read: such that) equal a particular property $P(x)$, which all elements of x satisfy, then we may simply write $\mathbb{N} = \{x: P(x)\}$.

It is also possible for a set to contain no elements. This memberless set is termed the empty or null set and is denoted by $\varnothing$. With the existence of this set given, we may proceed to develop a notational calculus for sets.

If A and B are two sets, and if every element of B is a member of A, we say that B is a *subset* of A, or A includes B, and write $B \subset A$ or, equivalently, $A \supset B$. This may be formalized as $B \subset A$ iff (read: if and only if) $x \in B \Rightarrow$ (read: implies that) $x \in A$ for all $x \in B$. When $B \subset A$ but $B \neq A$, B is termed a *proper* subset of A. And if $B \subset A$ and $A \subset B$, then $B = A$, that is, the sets A and B have the same elements and are therefore identical. Clearly, any set A has the obvious subsets A (i.e., $A \subset A$) and $\varnothing$ (i.e., $\varnothing \subset A$). The set of all possible subsets of A is termed the *power set* of A and is denoted by $\mathbb{P}(A)$. If two sets A and B have no elements in common, they are said to be disjoint. This is succinctly emphasized in terms of the intersection, which is a binary operation on sets.

Let A and B be two sets. The *union* of A and B is the set $A \cup B$, which contains those elements that are either in A or in B, or in both A and B. This may be formalized as $x \in A \cup B$ iff $x \in A$ and/or $x \in B$ or $A \cup B = \{x: x \in A \lor$ (read: disjunction) $x \in B\}$. The *intersection* of A and B is the set $A \cap B$, which contains only elements common to both A and B. Formally, $x \in A \cap B$ iff $x \in A$ and $x \in B$, or $A \cap B = \{x: x \in A \text{ and } x \in B\}$. In the case where A and B are disjoint, $A \cap B = \varnothing$. The proper *difference* of sets A and B is defined only when $B \subset A$. Here the difference of A and B is the set $A - B$, such that $x \in A - B$ iff $x \in A$ and $x \notin B$. The difference is often termed the relative complement of B in A.

An abstract set A is totally determined by its *cardinality*, denoted by $\#A$. For a finite set (i.e., a set with a finite number of elements), the cardinality is simply the finite cardinal number of elements in the set. The existence of infinite sets (sets with an infinite number of elements), however, leads to the concept of transfinite cardinal numbers, which we shall introduce only by example.

Consider once again the set of natural numbers $\mathbb{N} = \{1, 2, 3, \ldots\}$. This is the prototype *countable* infinite set, and its cardinality is the first transfinite cardinal number $\aleph_0$. Other infinite sets sharing this cardinality are the set of integers $\mathbb{Z} = \{\ldots, -2, -1, 0, 1, 2, \ldots\}$, and the set of rational numbers $\mathbb{Q} =$

$\{p/q : p, q \in \mathbb{Z}, q \neq 0\}$. The first *uncountable* infinite set has the transfinite cardinal number $\aleph_1$. The set of real numbers $\mathbb{R} = \mathbb{Q} \cup \{\text{irrationals}\}$ is a realization of this abstract set.

We define the sum of cardinal numbers as the cardinal number of a union of disjoint sets. The product of cardinal numbers is interpreted in terms of the *Cartesian product* of sets as follows. Let A and B be two sets with cardinality p and q, respectively. Then $A \times B$ designates the Cartesian product set, which consists of all ordered pairs or 2-tuples of the elements of A and B; the first member of each 2-tuple comes from A and the second from B. The cardinality of $A \times B$ is $p \cdot q$. As an example, let $A = \{a_1, a_2\}$, $B = \{b_1, b_2\}$. Then

$$A \times B = \{(a, b) a \in A, b \in B\} = \{(a_1, b_1), (a_1, b_2), (a_2, b_1), (a_2, b_2)\}$$

and the cardinality of $A \times B$ is $2 \cdot 2 = 4$. The countable extension is obvious and will not be formally presented.

2. Mappings

If A and B are sets, a function f from A to B is a *mapping* which assigns to each *preimage* element a in A a unique *image* element $f(a) = b$ in B. The elements of A constitute the *domain* of f while the *range* of f consists of those elements $b \in B$ for which there exists an $a \in A$ such that $f(a) = b$. This is interpreted as follows. The function f assumes the value b at the *argument* a; f then may be viewed as an operator which transforms or sends a into b. Symbolically we have $f : A \rightarrow B$, where f is a subset of the Cartesian product $A \times B$.

The mapping $f : A \rightarrow B$ may be formalized via the following properties:

1. f is a set and $f \subset A \times B$.
2. $f(A) \subset B$ defines A as the domain and guarantees that every point in A is defined in f. We say that every element of A is covered, or that every preimage must posses an image.
3. $\forall$ (read: for every) $a \in A \exists!$ (read: there exists uniquely) $f(a) = b \in B$. This statement embodies the "single-valued" nature of a mapping, that is, if a is fixed in A, there is only one value for $b \in B$ that satisfies the map f. This property ensures that no preimage may possess more than one image.
4. If $f(A) = B$, the image set B is covered and the map is said to be a *surjection* or an onto mapping.
5. If $\forall b \in B \exists! a \in A$ such that $f(a) = b$, every image has one and only one preimage and the map is said to be an *injection* or a one-to-one mapping.

6.   If a map is both injective and surjective, it is said to be a *bijection* or a
     one-to-one and onto mapping.

All maps considered here will be at least surjective.

3.   Groups

We begin our discussion of more complex set-based constructions by
considering the following structure axioms: Let S be a set and $f: S \times S \to S$ a
surjection. Then

A1. $\forall x, y, z \in S, \quad f(x, f(y, z)) = f(f(x, y), z)$
A2. $\forall x \in S, \quad \exists e \in S$ such that $f(e, x) = f(x, e) = x$
A3. $\forall x \in S, \quad \exists x' \in S$ such that $f(x', x) = f(x, x') = e$
A4. $\forall x, y \in S, \quad f(x, y) = f(y, x)$

(A1) ensures associativity of the binary combination rule which assigns to
every ordered pair in $S \times S$ an element in S. (A2) postulates the existence of an
identity element e, which is readily seen to be unique. (A3) states the existence
of an inverse element x' for every element of S, and these inverses are also
unique. (A4) describes the combination rule as being commutative. Notice that
closure is not specifically mentioned as an axiom here since it is implicit in the
definition of the binary operation in S as a surjective mapping from $S \times S \to S$.
    If we consider a set in conjunction with the structure axioms it satisfies, we
form a 2-tuple which is an object type. Symbolically we have $\mathscr{S} = (S, f)$, where
$\mathscr{S}$ is an object, S is a set, and f is the structure. We now examine several object
types in the order of their increasing complexity. If (A1) and (A2) are fulfilled,
the object $\mathscr{S}$ is called a *semigroup*. When $\mathscr{S}$ satisfies (A1), (A2), and (A3), a
group structure results. A group then may be defined as a set S and a binary
combination rule (group product) which is associative and for which an
identity and inverses exist. In general, group multiplication is not commutative. In the event that it is [i.e., in the event that (A4) is also satisfied], the group
is said to be *abelian*.
    Depending upon the context, the combination rule may be denoted as $+$,
indicating group "addition," with the identity being 0; or, the combination
rule may be denoted as $\cdot$ (normally suppressed), indicating group "multiplication," with the identity being 1. Whenever this notation is used, we define,
respectively, $+(x, y) \equiv x + y$ and $\cdot(x, y) \equiv x \cdot y$.
    Of particular interest to our impending development are the additive group
of real numbers $(\mathbb{R}, +)$, the multiplicative group of real numbers $(\mathbb{R}, \cdot)$, the
additive group of integers modulo n $(\mathbb{Z}_n, +)$, where n is a natural number, and
the multiplicative group of integers modulo p $(\mathbb{Z}'_p, \cdot)$, where p is prime. We

assume familiarity with the first two groups and discuss only the last two in detail.

ADDITIVE GROUP OF INTEGERS MODULO n $(\mathbb{Z}_n, +)$. We define congruence of two integers a and b, modulo the natural number n, by $a = b + kn$, where k is an integer. This may be written more concisely as $a = b \bmod n$. The *equivalence class* (or congruence class) of a modulo n is given by

$$[a]_n \equiv \{x \in \mathbb{Z} \mid x = a \bmod n\}$$

where $\mathbb{Z}$ is the set of integers. Simply put, $[a]_n$ is the set containing all integers which are congruent to a modulo n.

It is clear that every $a \in \mathbb{Z}$ is congruent modulo n to one of the numbers 0, 1, 2,..., $n - 1$. Now consider the set

$$\mathbb{Z}_n \equiv \{[0]_n, [1]_n, \ldots, [n - 1]_n\}$$

Under addition modulo n, $\mathbb{Z}_n$ is an additive abelian group of order n. (The order of a group is the cardinality of its underlying set.) The sum on $\mathbb{Z}_n$ is defined by

$$[a]_n + [b]_n = [a + b]_n$$

The identity element is $[0]_n$, and the inverse of an element $[a]_n$ is defined by

$$-[a]_n = [-a]_n = [n - a]_n$$

MULTIPLICATIVE GROUP OF INTEGERS MODULO p $(\mathbb{Z}'_p, \cdot)$. We define multiplication on $\mathbb{Z}_p$ by

$$[a]_p[b]_p = [ab]_p$$

Hence $[1]_p$ serves as the multiplicative identity. Also, since p is prime (see Ref. 2), every element has an inverse, excluding $[0]_p$. Thus the set

$$\mathbb{Z}'_p \equiv \{[1]_p, [2]_p, \ldots, [p - 1]_p\}$$

forms a group under multiplication modulo p. Furthermore, this group may be shown to be cyclic (i.e., $[2]_p^{p-1} \equiv [2^{p-1}]_p = [1]_p$).

4. Morphisms

When two objects contain additional structure, it is possible to enforce certain qualifications on functions between them to ensure that the operations of these objects are preserved. Such a map between two objects is called a *morphism*. Let

$$f: S \times S \to S \quad \text{and} \quad \mathscr{S}_1 \equiv (S, f)$$
$$g: S \times S \to S \quad \text{and} \quad \mathscr{S}_2 \equiv (S, g)$$
$$H: S \to S \quad \text{be a map with no associated structure}$$

We interpret the composite of maps f and H as follows:

$$f \circ H: S \times S \xrightarrow{H} S \times S \xrightarrow{f} S$$
$$H \circ f: S \times S \xrightarrow{f} S \xrightarrow{H} S$$
$$f \circ H(x, y) = f(H(x), H(y))$$
$$H \circ f(x, y) = H(f(x, y))$$

Diagramatically, we have

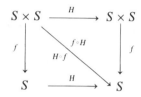

If $f \circ H = H \circ f$ holds pointwise, the above diagram is said to commute. H, then, is a surjective map which preserves structure and is called a *homomorphism*. Clearly, $f \circ H$ is a new structure, call it f', and $\mathscr{S}_1' \equiv (S, f \circ H \equiv f')$. Through map composition we have induced a way to get from $\mathscr{S}_1$ to $\mathscr{S}_1'$ by H, that is,

$$\mathscr{S}_1 \equiv (S, f) \xrightarrow{H} \mathscr{S}_1' \equiv (S, f')$$

and the objects $\mathscr{S}_1$ and $\mathscr{S}_1'$ are said to be homomorphic to each other.

Now we define the class of all homomorphisms of f, $\{H_i: i \in I\}$, where I is an index set, as

$$\text{Hom} f = \{f_i: f \circ H_i = f_i, \forall i \in I\}$$

Similarly, we define

$$\text{Hom} g = \{g_i: g \circ H_i = g_i, \forall i \in I\}$$

Given $f \in \mathrm{Hom}\, f$ and $g \in \mathrm{Hom}\, g$, if it happens that $\mathrm{Hom}\, f = \mathrm{Hom}\, g$, then

$$g \in \mathrm{Hom}\, f \Rightarrow g = f \circ H \text{ for some } H$$
$$f \in \mathrm{Hom}\, g \Rightarrow f = g \circ H' \text{ for some } H'$$

Therefore,

$$f = f \circ H \circ H' \Rightarrow H \circ H' = \mathscr{I}$$
$$g = g \circ H' \circ H \Rightarrow H' \circ H = \mathscr{I}$$

where $\mathscr{I}$ is the identity map.

This demonstrates that H possesses a point by point inverse, in this case H', and consequently may be characterized as a bijection. More appropriately, H is termed an *isomorphism*, that is, a bijective homomorphism. f and g are related to each other via H, and we say that f is isomorphic to g, written $f \sim g$. $f \sim g$ implies that the homomorphism classes of f and g are equal.

References

1. G. Birkhoff and S. MacLane, *A Survey of Modern Algebra*, rev. ed., Macmillan, New York, 1953.
2. A Clark, *Elements of Abstract Algebra*, Wadsworth, Belmont, CA, 1971.
3. P. R. Halmos, *Naive Set Theory*, Springer-Verlag Undergraduate Texts in Mathematics, Springer, New York, 1974.
4. F. Hausdorff, *Set Theory*, 2nd ed., Chelsea, New York, 1962.
5. A. P. Hillman and G. L. Alexanderson, *A First Undergraduate Course in Abstract Algebra*, 2nd ed., Wadsworth, Belmont, CA, 1978.
6. E. Kamke, *Theory of Sets*, Dover, New York, 1950.
7. I. Kaplansky, *Set Theory and Metric Spaces*, 2nd ed., Chelsea, New York, 1977.
8. H. J. Zassenhaus, *The Theory of Groups*, 2nd ed., Chelsea, New York, 1958.

4   The Genetic Code

Once the basic tenets of the central dogma of molecular genetics were formalized, much discussion ensued concerning the means by which the information encoded in the RNA message is interpreted during its translation into protein. Explicitly, how does the nucleotide sequence found in mRNA become deciphered and recast as the sequence of amino acids in a nascent polypeptide? Clearly, for this translation to occur, there must necessarily exist some well-defined rule or relationship between the informational units of these macromolecules. This relationship is embodied within the genetic code.

1.   Deciphering the Genetic Code

The genetic code is a rule that assigns every possible triplet (three-base) sequence of RNA bases to one of the 20 amino acids (or the termination operator) commonly incorporated into nascent polypeptide chains. That the genetic code must be at least a three-base code may be explained through a combinatorial argument alone. Since mRNA contains four nucleotide bases, a singlet code would determine only four amino acids while a doublet code would describe $4^2 = 16$ amino acids. Therefore a triplet code containing $4^3 = 64$ code words is the simplest possible code that can uniquely specify the necessary 20 amino acids. In addition, the genetic code interprets successive mRNA triplets with no punctuation or comas [1,2], that is, the mRNA codons are read sequentially without any signal to indicate where one codon ends and the next begins. Therefore the appropriate reading frame for mRNA translation (set by the initiation codon AUG) must be assumed, otherwise a polypeptide with an incorrect amino acid sequence will result.

The actual experiments to determine which of the 64 triplets specify each of the 20 amino acids required the confluence of genetic and biochemical techniques. Briefly, Nirenberg and Matthaei [3] incubated synthetic mRNAs with isolated ribosomes and other components of the protein-synthesizing machinery, including all 20 amino acids. They then determined which of the radioactively labeled amino acids had been incorporated into the resultant polypeptide. In this way, polyuridylic acid (poly U) and polycytidylic acid (poly C) were found to code for polymers containing only phenylalanine and proline residues, respectively. They therefore concluded that UUU is the code word for phenylalanine, while CCC codes for proline. Similarly, Ochoa and coworkers [4] determined that poly A codes for polylysine (i.e., AAA is the code word for lysine).

To investigate the coding relationship of the remaining triplets, synthetic copolymers in which the ratio of the two nucleotide bases is known were investigated. For example, in such a copolymer of poly UC (U:C ratio 5:1), the relative frequency of the eight possible triplets can be determined and compared with the amount of each amino acid incorporated into the resultant polypeptide to obtain putative codon: amino acid assignments (Table 4-1). Through such experiments [5], the empirical formula (but *not* the explicit sequence) of the code words for the individual amino acids was determined.

The next clue to the solution of the genetic code came from a binding assay developed by Nirenberg and Leder [6]. In this experiment, ribosomes, trinucleotides with known sequences, and aminoacyl tRNAs were incubated together and the specific binding of a trinucleotide (read in the $5' \to 3'$ direction) to an aminoacyl tRNA, in response to the tRNA's anticodon signal, was determined. In this way, the sequence of the nucleotide bases in the 61 triplets coding for the 20 amino acids was ascertained. The remaining three codons provide the termination signal for protein synthesis. Additional studies correlating the amino acid composition of polypeptides formed in response to repetitive polynucleotide messengers of known base sequence [7] provided

Table 4-1. Polypeptide Synthesis in Response to Poly UC (U:C Ratio 5:1)

Triplets	Triplet Frequency[a]	Amino Acid Incorporation[a]	
UUU	100	Phe	100
UUC ⎫ UCU ⎬ CUU ⎭	20	Phe ⎫ Ser ⎬ Leu ⎭	20
UCC ⎫ CUC ⎬ CCU ⎭	4	Ser ⎫ Leu ⎬ Pro ⎭	4
CCC	0.8	Pro	0.8

[a] Values are normalized for UUU (or Phe) = 100. Standard amino acid abbreviations are used (see Table 4-3).

corroborative evidence for the codon–amino acid assignments determined via the binding assays.

2. Characteristics of the Genetic Code

The coding relationship between mRNA triplet codons and amino acids may be presented in the form of a genetic code table (Table 4-2). This table is organized in such a way that several characteristics of the genetic code become immediately apparent. First, the genetic code is highly degenerate, that is, many amino acids are specified by more than one codon (Table 4-3). Codon degeneracy is not evenly distributed throughout the code, however. Some amino acids are sixfold degenerate (e.g., Arg, Leu, Ser), whereas other amino acids are only twofold degenerate (e.g., Asn, Asp). Of the 20 amino acids commonly incorporated into polypeptide chains during translation, only methionine and tryptophan are represented by single codons (i.e., these codons are nondegenerate).

Equally apparent from the genetic code table is the fact that the amino acids are not distributed randomly among the 64 triplets. As an example, all codons

Table 4-2. The Genetic Code[a]

UUU } Phe	UCU ⎫	UAU } Tyr	UGU } Cys
UUC	UCC ⎪ Ser	UAC	UGC
UUA } Leu	UCA ⎬	UAA } TC	UGA TC
UUG	UCG ⎭	UAG	UGG Trp
CUU ⎫	CCU ⎫	CAU } His	CGU ⎫
CUC ⎪ Leu	CCC ⎪ Pro	CAC	CGC ⎪ Arg
CUA ⎬	CCA ⎬	CAA } Gln	CGA ⎬
CUG ⎭	CCG ⎭	CAG	CGG ⎭
AUU ⎫	ACU ⎫	AAU } Asn	AGU } Ser
AUC ⎬ Ile	ACC ⎪ Thr	AAC	AGC
AUA ⎭	ACA ⎬	AAA } Lys	AGA } Arg
AUG Met	ACG ⎭	AAG	AGG
GUU ⎫	GCU ⎫	GAU } Asp	GGU ⎫
GUC ⎪ Val	GCC ⎪ Ala	GAC	GGC ⎪ Gly
GUA ⎬	GCA ⎬	GAA } Glu	GGA ⎬
GUG ⎭	GCG ⎭	GAG	GGG ⎭

[a] All codons are written in the $5' \to 3'$ direction from left to right. Standard amino acid abbreviations are given in Table 4-3; TC = termination codon.

Table 4-3. Degeneracy of the Genetic Code

Amino Acid	Three-Letter Abbreviation	Number of Codons
Alanine	Ala	4
Arginine	Arg	6
Asparagine	Asn	2
Aspartic acid	Asp	2
Cysteine	Cys	2
Glutamine	Gln	2
Glutamic acid	Glu	2
Glycine	Gly	4
Histidine	His	2
Isoleucine	Ile	3
Leucine	Leu	6
Lysine	Lys	2
Methionine	Met	1
Phenylalanine	Phe	2
Proline	Pro	4
Serine	Ser	6
Threonine	Thr	4
Tryptophan	Trp	1
Tyrosine	Tyr	2
Valine	Val	4

with a U in the second position code for hydrophobic amino acids. Also, acidic and basic amino acids are grouped together toward the bottom right-hand area of the table. The general impression is that related amino acids have to some extent related codons [8].

Clearly, the first two bases appear to be more significant for amino acid assignment than is the third base. In several cases XYN, where X and Y are fixed nucleotide bases and $N = $ U, C, A, G, codes for the same amino acid (e.g., $GUN = $ GUU, GUC, GUA, GUG $\mapsto$ Val). In the event that XYN defines two different amino acids, the third-base position in one case is occupied only by purines and in the other by pyrimidines (e.g., $GA_C^U = $ Asp, $GA_G^A = $ Glu). In general, then, amino acid codons occur in pairs and can be said to display a twofold degeneracy (XY_C^U or XY_G^A, or both). Exceptions to this basic symmetry are evidenced by the odd-order degenerate codons (e.g., Ile and TC with three codons each; Met and Trp with one codon each). We will return to the subject of the symmetry of genetic code degeneracies in Chapter 5.

On the basis of its somewhat less specific role in determining amino acid assignments, the third-base position of a codon is said to "wobble" [9]. According to the wobble hypothesis, the first two bases of a codon form strict base pairing with their anticodon counterparts. However, the first base in some anticodons (in the $5' \rightarrow 3'$ direction, i.e., the anticodon base that binds the third base of a codon) may interact with more than one codon for a specific amino acid. When the anticodon base in question is C or A, only one mRNA codon can be read. However, when the 5' anticodon base is U or G, two different codons may be interpreted. If the anticodon base is inosine (I, a modified nucleic acid base found in tRNAs), a total of three codons may be read. Schematically [10],

Anticodon $(3')X-Y-C(5')$ $(3')X-Y-A(5')$
                                          or
Codon $(5')Y-X-G(3')$ $(5')Y-X-U(3')$

Anticodon $(3')X-Y-U(5')$ $(3')X-Y-G(5')$
                                          or
Codon $(5')Y-X-^A_{G(wobble)}$ $(5')Y-X-^C_{U(wobble)}$

Anticodon $(3')X-Y-I(5')$

Codon $(5')Y-X-\overset{A}{\underset{C}{U}}$ (all wobble)

As a result of this wobble interaction, fewer tRNAs are required to translate the mRNA codons. (A minimum of 32 tRNAs are needed to interpret the 61 codons specifying the amino acids [10].)

Finally, the genetic code is traditionally thought to be universal, that is, the same codons indicate the same amino acids in all the cells of every living system. The universality of the genetic code has been explained by two different approaches. The stereochemical theory professes that the code is necessarily universal for stereochemical reasons alone. As an example, proponents of this theory contend that phenylalanine *has* to be represented by UU^U_C and by no other triplets, because in some way phenylalanine is stereochemically related to these two codons. Conversely, the frozen accident theory explains that the code is universal because any change would be lethal, or at least very strongly selected against. The rationale here is that since the code determines the amino acid sequences of several highly evolved protein molecules, any change in these molecules would be necessarily deleterious unless accompanied by concomitant mutations to rectify the mistakes produced by altering the code. While this accounts for the fact that the code does not change from organism to organism within a species, to account for the code being the same in all species one must assume that all life evolved from a single organism (or, strictly speaking, from a single closely breeding population) [11]. We will return to a more complete discussion of genetic code universality in Chapter 5, with the consideration of alternative or "ambiguous" amino acid assignments.

3. Origin and Evolution of the Genetic Code

Much speculation has occurred concerning the possible origins of the genetic code and the evolutionary development of this code to its present form [11–13]. As for the nature of a primitive code itself, two alternatives may be considered. (1) Suppose the primitive code was not a triplet code but that originally mRNA bases were read first one at a time (to yield four codons), then two at a time (for 16 codons), and finally three at a time for the present triplet code (with 64 codons). This idea of changing the codon size is basically unsatisfactory since such changes would necessarily make nonsense of *all* previous messages and would most likely be lethal. (2) Alternatively, it is more probable that the primitive code was also a triplet code, where perhaps the first two bases actually determined the amino acid coded for while the third position served as a punctuation signal.

The idea has also been advanced that the primordial code involved only a few amino acids since it is unlikely that all of the present amino acids were accessible at the time of the code's inception. Accordingly, the relatively rare amino acids (e.g., tryptophan and methionine) may be considered later additions to the code, while glycine, alanine, serine, and aspartic acid were presumably present initially [11]. The primordial amino acids may have effectively required only two of the three triplet bases for their specification (the third base serving merely as an intervening comma). However, the introduction of the more recent amino acids necessitated a fully functional three-base code. This third base is most easily understood as being derived from the signals used to punctuate the original code (Table 4-4). Supporting evidence for such ideas comes from the comparison of the amino acid composition of homologous proteins isolated from dissimilar species, for example ferredoxin from relatively "ancient" bacteria versus ferredoxin from present-day higher organisms [12]. In the former case, a small peptide containing only 13 types of amino acids is observed. However, the latter protein is a larger peptide in which wider varieties of amino acids are present.

Speculation concerning the nature of the primitive nucleic acids has also appeared. It is possible that primordial DNA contained only two bases A and I [11]. Adenine is thought to have been the base most likely available in the primitive environment, and inosine is derived from adenine by deamination. Presumably I coded in the same way as G does today (at least in the first two positions of the triplet). In this connection, note that present-day triplets containing only A or G in the first two positions (i.e., those found at the bottom right-hand corner of the code table) code for recognized primitive amino acids. The evolution of a primitive A, I nucleic acid to one possessing A, I, U, C must be accompanied by a change in the replication and transcription enzymes to recognize the smaller base pairs and a supply of the two new precursors (U and

Table 4-4. Possible Evolution of the Genetic Code[a]

Primitive Amino Acid	Primitive Codon[b]	Coding Recognition of Third Position	Present-Day Amino Acid
Lys	AAX	U, C	Asn
		A, G	Lys
Thr	ACX	U, C, A, G	Thr
Ser	AGX	U, C	Ser
		A, G	Arg
Ile	AUX	U, C, A	Ile
		G	Met
His	CAX	U, C	His
		A, G	Gln
Pro	CCX	U, C, A, G	Pro
Arg	CGX	U, C, A, G	Arg
Leu	CUX	U, C, A, G	Leu
Asp, Glu	GAX	U, C	Glu
		A, G	Asp
Ala	GCX	U, C, A, G	Ala
Gly	GGX	U, C, A, G	Gly
Val	GUX	U, C, A, G	Val
TC	UAX	U, C	Tyr
		A, G	TC
Ser	UCX	U, C, A, G	Ser
Cys	UGX	U, C	Cys
		A	TC
		G	Trp
Phe	UUX	U, C	Phe
		A, G	Leu

[a] After A. L. Lehninger, *Biochemistry*, Worth, New York, 1975.
[b] X indicates the presence of a punctuation signal in the third position of primitive codons.

C). Mutations would be effective in first producing the needed U and C in the chain. Eventually G would also be substituted for I to yield nucleic acids with the contemporary A, G, U, C bases. This type of an evolutionary scheme is particularly attractive since at no time would the message become complete nonsense.

In summary, then, it appears likely that in the evolution of the genetic code, only a few amino acids were coded initially. In the primitive situation, either the mechanism for coding was imprecise and could recognize most of the triplets, or only a few triplets were employed, possibly because the message initially contained effectively only two coding bases. In addition, it is highly probable that the recognition mechanism for code interpretation was not very reliable and a particular codon might correspond to a group of amino acids [13]. For example, consider the case when only the middle base is recognized. Under these conditions, U would code for a number of hydrophobic amino acids. In any event, it is generally believed that the primitive amino acids spread over the code until almost all of the triplets represented one or another of the amino acids [11]. Thus to avoid the existence of too many nonsense or termination triplets, most codons were quickly brought into use.

As the code evolved, there must necessarily have been an increase in the precision with which codons were recognized as well as an increase in the number of amino acids to be coded for. To meet this evolutionary challenge, the cell would have had to produce a new tRNA and activating enzyme in order to accommodate each new amino acid (or any minor amino acids already incorporated because of errors in recognition). This new tRNA might also have been called upon to recognize certain triplets already in use for the existing amino acids. In such a case, these triplets would now be ambiguous. For this sequence of changes to be finalized, further mutations must have occurred which resulted in the replacement of the ambiguous codons with other triplets. Eventually codons would no longer be ambiguous and would code only for the new amino acids. This mechanism for the introduction of new amino acids could have been successful only if the primitive genetic message coded for a small number of crudely constructed proteins. As the evolution of the genetic code proceeded, more "sophisticated" proteins would have been produced until no new amino acids could have been introduced without seriously affecting large numbers of proteins. At this point the code is said to be frozen. The evolutionary scheme described above does not dictate that the original codons of the primitive code must maintain their primordial amino acid assignments. Therefore the evolution of the code may have obliterated all traces of the primitive code [11].

References

1. F. H. C. Crick, L. Barnett, S. Brenner, and R. J. Watts-Tobin, *Nature*, **192**, 1227 (1961).
2. F. H. C. Crick, *Sci. Am.*, **215** (4) 55 (1966).
3. M. W. Nirenberg and J. H. Matthaei, *Proc. Natl. Acad. Sci. USA*, **47**, 1588 (1961).

4. P. Lengyl, J. F. Speyer, and S. Ochoa, *Proc. Natl. Acad. Sci. USA*, **47**, 1936 (1961).

5. M. Nirenberg, *Sci. Am.*, **208** (3) 80 (1963).

6. M. W. Nirenberg and P. Leder, *Science*, **145**, 1399 (1964).

7. H. P. Ghosh, D. Söll, and H. G. Khorana, *J. Mol. Biol.*, **25**, 275 (1967).

8. C. J. Epstein, *Nature*, **210**, 25 (1966).

9. F. H. C. Crick, *J. Mol. Biol.*, **19**, 548 (1966).

10. A. L. Lehninger, *Principles of Biochemistry*, Worth, New York, 1982.

11. F. H. C. Crick, *J. Mol. Biol.*, **38**, 367 (1968).

12. T. H. Jukes, *Molecules and Evolution*, Columbia University Press, New York, 1966.

13. C. R. Woese, *The Genetic Code*, Harper & Row, New York, 1967.

5 Genetic Code Symmetries

1. Introduction

Through the processes of transcription and translation, the information contained in the genetic material of cells determines the production of newly synthesized protein molecules. During transcription, DNA provides the necessary template for the production of mRNA. The message encoded in these mRNA molecules is subsequently translated into protein. The subject of this chapter concerns a mathematical explication of the genetic coding process which operates at the level of translation.

The genetic code may be conceptualized as a language, grounded on the four nucleic acid bases found in mRNA, through which 64 triplet (three-base) codons enumerate the 20 common amino acids and the terminator codons (TC). Clearly, the genetic code is highly degenerate (i.e., most amino acids are specified by more that one codon). However, these degeneracies are not randomly dispersed throughout the genetic code table. Rather, the genetic code possesses a definite structure which is evidenced in the regular, symmetric pattern of codon degeneracies [1–4]. In addition, careful determinations of complementary mRNA and protein sequences indicate that significant deviations from standard genetic codon assignments often exist. Such "ambiguous" or, more appropriately, "alternative" codings are traditionally regarded as aberrant occurrences which simply indicate that the protein synthesis system is not without error.

The objectives of this chapter are then twofold. First, we investigate the inherent symmetry of codon degeneracies in the standard genetic code (SGC) from the viewpoint of the set- and group-theoretic concepts detailed in Chapter 3. Such an investigation provides an extension of the SGC to embrace alternative codon assignments within a generalized genetic code (GGC) [5–11]. Second, we provide an extensive examination of the biological context within which translation proceeds, as a necessary means for unifying alternative codings with those of their SGC counterparts, thereby reestablishing the universality of the genetic code within the framework of the GGC.

We stress from the outset that our approach to genetic coding theory is fundamentally different from the standard treatment. In particular, the thesis we explore implicitly invokes multiple genetic codes as a means of reconciling the ever-increasing assembly of alternative codon assignments.

2.   Basic Symmetry of the Standard Genetic Code

The genetic coding process, viewed abstractly, is based on the set of four ribonucleic acid bases $B \equiv \{U, C, A, G\}$. The elements of B ($U \equiv$ uracil, $C \equiv$ cytosine, $A \equiv$ adenine, $G \equiv$ guanine) are the pyrimidine (U, C) and purine (A, G) bases which control the functionality of an mRNA message. The linear sequence of these bases in a particular mRNA template determines the corresponding linear sequence of amino acids in the resultant protein molecule. Each triplet (or *codon*) of base elements provides for the introduction of a specific amino acid into a nascent polypeptide.

The set of codons C contains all possible 3-tuples of the set B, that is, the relationship between B and C may be expressed by the Cartesian product

$$C = B \times B \times B \tag{5-1}$$

Thus C includes $4^3 = 64$ codons. A third set A consists of the 20 common amino acids and the TC operator. The usual biological reasoning describes

Table 5-1. Standard Genetic Code[a]

UUU UUC	Phe	UCU UCC		UAU UAC	Tyr	UGU UGC	Cys
UUA UUG	Leu	UCA UCG	Ser	UAA UAG	TC	UGA UGG	TC Trp
CUU CUC CUA CUG	Leu	CCU CCC CCA CCG	Pro	CAU CAC	His	CGU CGC CGA CGG	Arg
				CAA CAG	Gln		
AUU AUC AUA	Ile	ACU ACC ACA	Thr	AAU AAC	Asn	AGU AGC	Ser
AUG	Met	ACG		AAA AAG	Lys	AGA AGG	Arg
GUU GUC GUA GUG	Val	GCU GCC GCA GCG	Ala	GAU GAC	Asp	GGU GGC GGA GGG	Gly
				GAA GAG	Glu		

[a] Each codon is written in the $5' \rightarrow 3'$ direction from left to right. The amino acid (or terminator codon) to which each codon corresponds is written to the right of the codon. Standard abbreviations for the amino acids are used.

the relationship of C with A as a mapping

$$f: C \rightarrow A \qquad (5\text{-}2)$$

This mapping has been determined by chemical or genetic methods and is presented explicitly in Table 5-1. Equation (5-2) specifies what we refer to as the standard genetic code (SGC) [5–11]. As conventionally formalized, the SGC demands that f map C onto A. The surjective (onto) nature of this mapping delimits the degeneracy of the SGC, that is, an amino acid may be specified by more than one codon; or, rephrased set-theoretically, an image (an element of set A) may have more than one preimage (elements of set C).

In the event that more than one amino acid is specified by the same codon, the fidelity of the SGC is compromised. Such ambiguous codon assignments violate the necessary single-valued nature of a mapping (i.e., no preimage may possess more than one image), and indicate that the genetic code should be envisioned as a relation rather than as a mapping. Thus the essential point here is that a relation may be multivalued. The distinction between a mapping and a relation and their significance for genetic coding theory are illustrated diagrammatically in Figure 5-1.

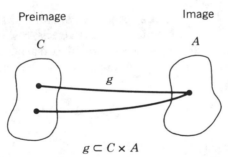

Surjective (onto) mapping ($\equiv$ degeneracy)

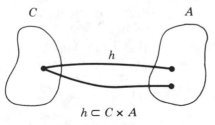

Multivalued relation ($\equiv$ ambiguity)

Figure 5-1. Distinction between a mapping and a relation using set diagrams.

Gatlin [12] has suggested that the intrinsic ambiguous nature of a coding relation may be lifted through the introduction of a specific biological context within which genetic code interpretation proceeds. By way of example, consider cytoplasmic protein synthesis initiation. It is well known that AUG and GUG serve as initiators of protein synthesis but code for methionine and valine, respectively, when situated downstream (posterior) to the initiator site on an mRNA molecule. Therefore, in such a case we may distinguish initiation as a specific biological context. Whereas ambiguity during initiation is generally recognized, the prevalence of ambiguity internal to the mRNA message is conventionally questioned. Instead, the traditional wisdom sustains a rigorous genetic code mapping and excuses any ambiguities as mere translational errors. Although this viewpoint could be justifiable for in vitro ambiguities, it is considerably more questionable in light of the accumulated in vivo data [7, 10–12]. Therefore, we proceed on the assumption of the reality of ambiguity and explore its consequences, particularly with respect to the introduction of biological contexts as integral facets of the genetic code.

With the supposition of ambiguity fully admitted, the genetic code must now be characterized by the relation R, where

$$R \subset C \times A \tag{5-3}$$

R is a proper subset of $C \times A$ [5, 6, 8]. A particular biological context selects a unique subset of R for which the mapping characteristic is again established. The presence of coding ambiguities necessitates the existence of more than one context and, therefore, more than one mapping. Symbolically, one has

$$(R, C_i) = f_i, \qquad i \in I \tag{5-4}$$

where

$$f_i : C \to A \tag{5-5}$$

In Equation (5-4) C_i designates a specific biological context, I is some finite index set, and (R, C_i) indicates the selection of a particular mapping f_i from R under the ith biological context. For example, when C_1 corresponds to normal cytoplasmic protein synthesis, then $(R, C_1) = f_1 \equiv f$. The GGC may therefore by defined as the class of all such contextual mappings

$$\{f_i, i \in I \mid f_i : C \to A\} \tag{5-6}$$

Consequently, a biological context restricts the GGC to one of many specific genetic codes and thereby confines the GGC relation to a fixed mapping.

Operationally, the context structure reduces the question of the possible universality of the *SGC* to that of the possible universality of the *SGC* context.

3. *SGC* and *GGC* Symmetries

A. *Symmetry of the SGC*

The inherent symmetry characteristics of the *SGC* have been argued extensively ([1–4] and Chapter 4). Briefly, such symmetries are apparent from a careful inspection of the genetic code (Table 5-1). Clearly, for almost all codons, a substitution in the third nucleotide position between purine $(A \rightleftharpoons G)$ or pyrimidine $(U \rightleftharpoons C)$ bases does not affect the amino acid assignment. To formalize the *SGC* codon degeneracies, it is necessary to introduce a decomposition D_0 of C. We define D_0 as $\{C_k, k \in B\}$ where

$$C_k = \{(i, j, k) \in C \mid i, j \in B\} \tag{5-7}$$

where i, j, k designate the first, second, and third codon bases, respectively, and $\bigcup_{k \in B} C_k = C$. D_0 partitions C into four subsets (of 16 elements each), each of which contains only those codons having the same third base (Table 5-2). It is clear that

$$f(C_U) = f(C_C), \qquad f(C_A) \neq f(C_G) \tag{5-8}$$

The inequality is the result of odd-order degenerate codons for isoleucine, tryptophan, methionine, and the termination operator TC.

Consequently, Table 5-2 indicates a simple relationship for the even-order degenerate codons. With D_0 we may analyze *SGC* codon degeneracies given the following two sets:

$$F_1 = \{(\alpha, \beta) \in C \times C \mid f(\alpha) \in f(C_U), f(\beta) \in f(C_C)\} \tag{5-9a}$$

$$F_2 = \{(\alpha, \beta) \in C \times C \mid f(\alpha) \in f(C_A), f(\beta) \in f(C_G)\} \tag{5-9b}$$

The sets F_1 and F_2 are mutually exclusive for a doubly degenerate codon pair. The importance of the existence of these sets is that, given the C_k location of one member of a codon pair, we have a rule that assigns a unique location to the other member. In other words, F_1 and F_2 define a one-to-one correspondence from one member of a doubly degenerate codon pair to the other member.

Table 5-2. Decomposition of C Induced by $C_k \equiv \{(i,j,k)\in C \mid i,j\in B\}$

C_U	$f(C_U)$	C_C	$f(C_C)$	C_A	$f(C_A)$	C_G	$f(C_G)$
UCU ⎫		UCC ⎫		UCA	Ser	UCG	Ser
AGU ⎬	Ser	AGC ⎬	Ser	AGA ⎫		AGG ⎫	
CGU	Arg	CGC	Arg	CGA ⎬	Arg	CGG ⎬	Arg
CUU	Leu	CUC	Leu	CUA ⎫		CUG ⎫	
GCU	Ala	GCC	Ala	UUA ⎬	Leu	UUG ⎬	Leu
GUU	Val	GUC	Val	GCA	Ala	GCG	Ala
CCU	Pro	CCC	Pro	GUA	Val	GUG	Val
GGU	Gly	GGC	Gly	CCA	Pro	CCG	Pro
ACU	Thr	ACC	Thr	GGA	Gly	GGG	Gly
UUU	Phe	UUC	Phe	ACA	Thr	ACG	Thr
UAU	Tyr	UAC	Tyr	CAA	Gln	CAG	Gln
UGU	Cys	UGC	Cys	AAA	Lys	AAG	Lys
CAU	His	CAC	His	GAA	Glu	GAG	Glu
GAU	Asp	GAC	Asp	UAA ⎫		UAG	TC
AAU	Asn	AAC	Asn	UGA ⎬	TC	UGG	Trp
AUU	Ile	AUC	Ile	AUA	Ile	AUG	Met

The analogous characterization for order-4 degeneracies is completed by allowing F_1 and F_2 to hold simultaneously. Hence as shown in Table 5-2, one and only one codon from an order-4 degenerate codon set must list in every $f(C_k)$ column. The extension to a degeneracy of order 6 is straightforward. In the case of four of the six codons, both F_1 and F_2 hold, whereas for the last two codons, either F_1 or F_2 holds.

The exact nature of the one-to-one correspondence defined by F_1 and F_2 follows from the decomposition D_0. For any codon pair (from an even-order degenerate codon set) that is an element of either F_1 or F_2, the first two bases of one codon of the pair are identical to those of the other codon. The third base, on the other hand, is different for the two codons but must be pyrimidine (i.e., for F_1) or purine (i.e., for F_2). This, of course, is the inherent symmetry of the f mapping, and we will refer to it as the even-order degeneracy constraint.

In summary, then, even-order (2, 4, or 6) codon degeneracies obey the following necessary (but not sufficient) conditions.

Order 2. Two codons are doubly degenerate if they form a codon pair which is an element of *either* F_1 or F_2.

Order 4. Four codons are fourfold degenerate if two of the codons form a codon pair that is an element of F_1 *and* the other two codons form a codon pair that is an element of F_2.

Order 6. Six codons are sixfold degenerate if two of the codons form a codon pair that is an element of F_1 *and* two more of the codons form a codon pair that is an element of F_2; the remaining two codons then form a codon pair that is an element of *either* F_1 *or* F_2.

If the *SGC* is supposed to be exact, odd-order codon degeneracies (including those codons that are nondegenerate, such as the single codons specifying the amino acids tryptophan and methionine) suggest that an evolutionary change of the genetic code has produced slight deviations from the symmetry noted above. Consider the two sets of triply degenerate codons, one of which maps onto Ile while the other maps onto TC. Two of the codons which correspond to Ile satisfy the even-order degeneracy constraint and F_1. Thus these two codons may be characterized as doubly degenerate. The third codon is AUA, and we denote $f(AUA) = Ile^*$. The case of TC is more unique: either of the codon pairs (UAA, UAG) or (UGA, UAG) satisfies F_2. However, the first pair also satisfies the even-order degeneracy constraint. Therefore (UAA, UAG) may be characterized as a doubly degenerate codon pair. The third codon is UGA, and we designate $f(UGA) = TC^*$.

Consider, finally, the two nondegenerate codons, one of which maps onto Met (i.e., AUG) while the other maps onto Trp (i.e., UGG). According to the preceding discussion, however, the codons corresponding to Ile* and TC* may also be considered to be effectively nondegenerate. Allowing the amino acid symbols to represent the actual codons in question, we find from Table 5-2 that the codon pairs (TC*, Trp) and (Ile*, Met) satisfy the even-order degeneracy constraint. Thus one may infer that these codon pairs are doubly degenerate. In fact, a number of both in vivo and in vitro codon assignments (see Tables 5-6 and 5-7) indicate that, under certain conditions, the codon pairs (Ile*, Met) and (TC*, Trp) are doubly degenerate and map onto the amino acids Met and Trp, respectively.

Therefore for the odd-order degenerate codons, the even-order degeneracy constraint indicates that slight modifications of this strict symmetry have occurred. However, it is possible to amplify and generalize the concept of symmetry deviations. For example, one may assume the converse stance to that taken above, that is, taking the symmetry of codon degeneracies to be exact while treating the *SGC* as nonexact. This leads to the *GGC*.

B. Symmetry of the GGC

Codon degeneracies under the *GGC* may be considered given the following two sets:

$$R_1 = \{(\alpha, \beta) \in C \times C \mid \alpha \in C_U, \beta \in C_C\} \qquad (5\text{-}10a)$$

$$R_2 = \{(\alpha, \beta) \in C \times C \mid \alpha \in C_A, \beta \in C_G\} \qquad (5\text{-}10b)$$

Obviously, R_1 and R_2 are derived from F_1 and F_2 by the removal of the mapping characteristic from the latter sets. One may now consider GGC codon degeneracies in the same manner as the SGC, with F_1, F_2 replaced by R_1, R_2. In fact, the basic symmetry of the SGC codon degeneracies is indeed encompassed by the GGC since $F_1 \subset R_1$ and $F_2 \subset R_2$.

While D_0 adequately describes SGC degeneracies and ambiguous codings resulting from third-base wobble [13], certain alternative codings require the introduction of additional decompositions of the codon set C which have symmetry characteristics similar to D_0 under the SGC f mapping and which permit the construction of sets F_1 and F_2. Seven other decompositions, D_1–D_7, may be devised through group-theoretic construction [5–11].

Imposition of a group structure on the set of nucleotide bases, with concomitant treatment of the Cartesian product as group multiplication, requires C to become an assemblage of triple products of the elements of B. Furthermore, C is equivalent to B when the repeated elements are removed so as to again form a set. Thus when B is a group,

$$B = B \times B \times B \tag{5-11}$$

The number of elements of C which map onto one of the group elements of B is given as the cardinality of C divided by the order of B (i.e., $64/4 = 16$). The pairwise disjoint subsets of C are designated C_k, $k \in B$, where the subscripts denote the elements of B that the codons of the subset map onto under group multiplication. (This notation is analogous to that presented in Section 3.A and will be justified below. It must be remembered, however, that the actual elements of a subset C_k as defined above are different from those of Section 3.A.)

Now since B must be a group of order 4, there are only two abstract possibilities, both of which are abelian: the Klein four group V, and the cyclic group of order four Z_4. The abstract presentation of V is*

$$V: \langle a, b; a^2, b^2, ab = ba \rangle \tag{5-12}$$

and the group elements are $V = \{1, a, b, c (\equiv ab)\}$, where 1 is the identity element. Z_4 is presented as

$$Z_4: \langle d; d^4 \rangle \tag{5-13}$$

---

*In the formalism of an abstract presentation of a group, elements to the left of the semicolon are the group generators, while expressions to the right of the semicolon are either relators or relations. See, for example, [14].

Table 5-3. Distinct Order Isomorphisms $B \leftrightarrow V$, Z_4

$B \leftrightarrow V$		$B \leftrightarrow Z_4$	
Label	Order Isomorphism	Label	Order Isomorphism
V-1	$\{U, A, C, G\}$	Z_4-1	$\{U, A, C, G\}$
V-2	$\{C, G, U, A\}$	Z_4-2	$\{C, G, U, A\}$
V-3	$\{A, C, G, U\}$	Z_4-3	$\{A, C, G, U\}$
V-4	$\{G, U, A, C\}$	Z_4-4	$\{G, U, A, C\}$
		Z_4-5	$\{U, A, G, C\}$
		Z_4-6	$\{G, C, U, A\}$
		Z_4-7	$\{A, G, C, U\}$
		Z_4-8	$\{C, U, A, G\}$
		Z_4-9	$\{U, C, A, G\}$
		Z_4-10	$\{A, G, U, C\}$
		Z_4-11	$\{C, U, G, A\}$
		Z_4-12	$\{G, A, C, U\}$

and $Z_4 = \{1, d, e\, (\equiv d^2), f\, (\equiv d^3)\}$. To arrive at decompositions of C, one makes an order isomorphism from B to either V or Z_4. In general, there are 4!, or 24, order isomorphisms for each abstract group, for a total of 48. Nevertheless, the (tedious!) job of constructing multiplication tables shows that there are only four *distinct* order isomorphisms for V and 12 for Z_4 (see Table 5-3 and Section 4). When these distinct order isomorphisms are used to induce C decompositions, however, one finds that there are only seven distinct decompositions, D_1–D_7. These are presented in Table 5-4.

We are now in a position to explore the nature of D_1–D_7 under the f mapping. Proceeding in a manner analogous to that of Section 3.A, we find that there exist three sets of one-to-one correspondences, L_1–L_3, for the D_1–D_7 decompositions. These are defined by the doubly degenerate codons (α, β) in the following manner.

L_1, applicable to D_1, D_2, D_3, either

1. $f(d_1^{(a,2)}) \in f(C_U) \Leftrightarrow f(d_2^{(a,2)}) \in f(C_C)$

or

2. $f(d_1^{(a,2)}) \in f(C_A) \Leftrightarrow f(d_2^{(a,2)}) \in f(C_G)$

L_2, applicable to D_4, D_5, either

1. $f(d_1^{(a,2)}) \in f(C_U) \Leftrightarrow f(d_2^{(a,2)}) \in f(C_C)$

Table 5-4. Group-Theoretic Decompositions of C^a

Label	Order Isomorphisms	Subsets of C			
		C_U	C_C	C_A	C_G
D_1	V-1, V-2, V-3, V-4	CAG	UAG	GUC	AUC
		UCC	CUU	AGG	GAA
		UAA	CGG	AUU	GCC
		UGG	CAA	ACC	GUU
		UUU	CCC	AAA	GGG
D_2	Z_4-1, Z_4-2	UAG	CAG	GUC	AUC
		UCC	CUU	GAA	AGG
		CGG	UAA	AUU	GCC
		CAA	UGG	ACC	GUU
		UUU	CCC	GGG	AAA
D_3	Z_4-3, Z_4-4	CAG	UAG	AUC	GUC
		CUU	UCC	AGG	GAA
		UAA	CGG	GCC	AUU
		UGG	CAA	GUU	ACC
		CCC	UUU	AAA	GGG
D_4	Z_4-5, Z_4-6	AUC	UAG	GUC	CAG
		GAA	CUU	AGG	UCC
		GCC	CGG	AUU	UAA
		UGG	ACC	CAA	GUU
		UUU	AAA	CCC	GGG
D_5	Z_4-7, Z_4-8	CAG	GUC	UAG	AUC
		UCC	AGG	CUU	GAA
		UAA	AUU	CGG	GCC
		GUU	CAA	ACC	UGG
		GGG	CCC	AAA	UUU
D_6	Z_4-9, Z_4-10	GUC	UAG	CAG	AUC
		AGG	CUU	UCC	GAA
		UAA	GCC	AUU	CGG
		ACC	CAA	UGG	GUU
		UUU	GGG	AAA	CCC
D_7	Z_4-11, Z_4-12	CAG	AUC	GUC	UAG
		UCC	GAA	AGG	CUU
		AUU	CGG	UAA	GCC
		UGG	GUU	ACC	CAA
		AAA	CCC	UUU	GGG

^{a}Each codon entry represents all permutations of its three bases. For example, CAG represents CAG, AGC, GCA, CGA, ACG, and GAC.

or

   2. $f(d_1^{(a,2)}) \in f(C_A) \Leftrightarrow f(d_2^{(a,2)}) \in f(C_G)$

or

   3. $f(d_1^{(a,2)}) \in f(C_U) \Leftrightarrow f(d_2^{(a,2)}) \in f(C_A)$

or

   4. $f(d_1^{(a,2)}) \in f(C_C) \Leftrightarrow f(d_2^{(a,2)}) \in f(C_G)$

L_3, applicable to D_6, D_7, either

   1. $f(d_1^{(a,2)}) \in f(C_U) \Leftrightarrow f(d_2^{(a,2)}) \in f(C_C)$

or

   2. $f(d_1^{(a,2)}) \in f(C_A) \Leftrightarrow f(d_2^{(a,2)}) \in f(C_G)$

or

   3. $f(d_1^{(a,2)}) \in f(C_U) \Leftrightarrow f(d_2^{(a,2)}) \in f(C_G)$

or

   4. $f(d_1^{(a,2)}) \in f(C_C) \Leftrightarrow f(d_2^{(a,2)}) \in f(C_A)$

L_1, of course, is the same one-to-one correspondence as that which generates F_1, F_2. Thus the discussion and the results of Section 3.A for the decomposition D_0 are equally valid for the decompositions D_1, D_2, and D_3.

Obviously, L_1 is weaker than L_2 or L_3. In fact, L_1 is itself contained within L_2 or L_3. The difficulty in discussing the symmetry of the f mapping with L_2 or L_3 is that the even-order degeneracy constraint is not readily apparent for even-order degeneracies greater than 2 when these implications are used. This can lead to ambiguities in the application of L_2 or L_3 to the order-4 and order-6 degenerate codon sets. However, the even-order degeneracy constraint is clear for L_2 and L_3 when the order-2 degeneracies are considered, and this, of course, is what induces these sets of one-to-one correspondences. Thus if the even-order degeneracy constraint is postulated for the treatment of order-4 and order-6 degeneracies by L_2, L_3, all ambiguity is alleviated. It then follows that an order-4 degeneracy is specified uniquely by allowing two of the L_2 (or

L_3) implications to hold simultaneously. Likewise, an order-6 degeneracy is specified uniquely by allowing three of the L_2 (or L_3) implications to hold simultaneously.

Up to this point we have treated the genetic code as a defined mapping. This, in turn, induces the one-to-one correspondence L_1 for D_0–D_3. In order to discuss ambiguous codings, however, it is necessary to treat the genetic code as a relation, that is, we consider the genetic code to be a subset of $C \times A$, which lacks the mapping characteristic. Our previous line of reasoning leads us to postulate that, under conditions when the genetic code must be treated as a relation, L_1 must also be treated as a relation. Thus we consider the underlying symmetry discussed above to be fundamental to the biological process of code interpretation within an appropriate biological context. The generalization of this symmetry results from the removal of the mapping characteristic. This, then, leads us to postulate the specific relations R_1, R_2 [see Eqs. (5-10a, b)] which serve as the necessary (but not sufficient) conditions for codon degeneracy. The content of this postulate for an order-2 degenerate codon set is as follows. If a codon pair is doubly degenerate, then it must be an element of either R_1 or R_2. The validity of this statement, of course, relies totally on how well it reflects known codon assignments (see Section 5).

For an order-4 degenerate codon set, the necessary (but not sufficient) conditions that the four codons be four-fold degenerate is that two of the codons form a codon pair that is an element of R_1 and the other two codons form a codon pair that is an element of R_2. In a similar manner, for an order-6 degenerate codon set, the necessary (but not sufficient) conditions that the six codons be sixfold degenerate is that two of the codons form a codon pair that is an element of R_1 and two more of the codons form a codon pair that is an element of R_2; the last two codons must then form a codon pair that is an element of either R_1 or R_2.

It is obvious, of course, that R_1 and R_2 derive from $L_1 - 1$ and $L_1 - 2$, respectively, by removal of the mapping characteristic. However, the use of R_1 and R_2 implies that a particular C decomposition has been chosen. Thus there are actually eight sets of relations, one each for D_0–D_7. It is important to note, however, that if the sets R_1 and R_2 are formed for all of the decompositions D_0–D_7, not all of the 64^2 possible codon pairs will be represented. This obviously must be the case, or otherwise the generalization would have no content: That is, if all codon pairs could be an element of some R_1 or R_2, necessity would be trivially satisfied.

The reason for basing the generalization on L_1 is twofold. First, we wished to construct a relation from which the symmetry of f could be reconstructed as simply as possible. Such a result naturally obtains if the generalization is from L_1 rather than from L_2 or L_3. Second, relations formed from L_2 or L_3 are greater in cardinality than those formed from L_1. An increase in cardinality, in turn, strengthens the necessity condition for codon degeneracy. It should be noted also that if relations formed from L_2 and L_3 are used simultaneously, then necessity is satisfied trivially, as discussed above. Thus we have chosen the weakest set of relations from the three that are possible and apply them to all decompositions D_0–D_7. In the end, the utility of D_4–D_7 is evidenced by the ambiguous codon assignments, and these codings also provide a test of the necessity postulate, R_1 and R_2. We will return to a detailed discussion of these data in Section 5.

4. Order Isomorphism/Order Equivalence Relations

It is now convenient to summarize and further discuss the details of the genetic code symmetry analysis of the preceding sections. As a result of these analyses, we have shown that the GGC satisfies a (weak) symmetry principle which incorporates, and generalizes, the degeneracy pattern of the SGC. Clearly, this symmetry principle devolves on the specification of necessary (but not sufficient) conditions for codon degeneracy and serves as a constraint on admissible f_i codes. The nature of the constraint was first elucidated from the SGC degeneracies and then extended to the GGC via the introduction of group-theoretically induced partitions of the codon set C. The technique employed was to define order isomorphisms from the RNA base set B to both the cyclic group of order 4, Z_4, and the Klein four group V. This gives a total of 48 order isomorphisms, not all of which are distinct. (The reader will recall that 12 are distinct for Z_4 but only four are distinct for V.) The distinct order isomorphisms (technically, order equivalence classes, see below) are then used to generate partitions of C by treating $B \times B \times B$ as group multiplication and collecting into $C_k \subset C$ all codons mapping onto $k \in B$ under the group product. In this way, seven partitions were generated, one for V and six for Z_4. And, finally, these partitions are used to analyze the degeneracy symmetries of alternative codings in a manner completely analogous to that for the SGC.

There remain several unsolved problems in genetic coding theory which are concerned, in principle, with the following two desiderata.

1. It is the ultimate goal of this approach to genetic coding theory to explicitly specify all possible alternative codes f_i statifying the generalized codon degeneracy symmetry discussed above.
2. It is desirable to be able to extend the above theory to genetic codes employing an arbitrary number of bases per codon. This would be a necessary prerequisite to address the problem of evolutionary development of degeneracy symmetries within a recently proposed [15] five-base codon model for a primitive genetic code.

Unfortunately, progress on both of these problems (of which the first is undoubtedly the more important) has been impeded by our very imperfect understanding of distinct order isomorphism/partition correlations. The purpose of this section is to provide a first step in the direction of such an understanding.

In our earlier discussion, nondistinct order isomorphisms were verified by direct computation. (This laborious effort could be justified initially since the only groups that required investigation, for the biological problem at least, were the two groups of order 4.) However, the insight necessary for specifying all admissible f_i's requires a full understanding of the *equivalence* of order isomorphisms for groups of arbitrary order. This is provided below, albeit in a form necessarily divorced from our original application. Our principal combinatorial result is as follows. If G is a group of order n, the number of distinct order isomorphisms mapping G to G is $n!/m$, where m is the number of automorphisms of G.

ORDER ISOMORPHISMS AND ORDER EQUIVALENCE CLASSES.   Let S be a set of cardinality n (i.e., $\#S = n$) and consider the class of all bijections h_i on S,

$$h_i : S \to S, \qquad i = 1, \ldots, n! \tag{5-14}$$

The set $\{h_i\}$, under map composition, forms the symmetric group on n objects, S_n. [In what follows, we shall take $h_i : S \times S \to S \times S$ to be defined by $(x, y) \mapsto (h_i(x), h_i(y))$ for all $(x, y) \in S \times S$.] Given an appropriate structure map

$$g : S \times S \to S \tag{5-15}$$

let $G := (S, g)$ be a group of order n.

We define a *transformed* structure map $g_i := h_i^{-1} \cdot g \cdot h_i$ in terms of the following commutative diagram:

$$
\begin{array}{ccc}
S \times S & \xrightarrow{\;h_i\;} & S \times S \\
{\scriptstyle g_i}\Big\downarrow & & \Big\downarrow{\scriptstyle g} \\
S & \xleftarrow[\;h_i^{-1}\;]{} & S
\end{array}
\tag{5-16}
$$

Now h_i is a homomorphism (and, hence, an isomorphism) if and only if, for all $x, y \in S$,

$$h_i(g(x, y)) = g(h_i(x), h_i(y))$$

or, more conveniently,

$$h_i \cdot g = g \cdot h_i \tag{5-17}$$

Thus h_i is an isomorphism (technically, an *automorphism*) if and only if

$$g = h_i^{-1} \cdot g \cdot h_i =: g_i \tag{5-18}$$

Define the equivalence class

$$[g] := \{g_i, i = 1, \ldots, n!: g_i = g\}$$

and, in general,

$$[g_i] := \{g_j, j = 1, \ldots, n!: g_i = g_j\} \tag{5-19}$$

Finally, define

$$H = \{h_i, i = 1, \ldots, n!: g_i = g\} \tag{5-20}$$

REMARKS

1. h_i is an *order isomorphism*, but h_i is an isomorphism in the usual sense (i.e., an automorphism) if and only if $g_i \in [g]$.
2. It is trivial to show that H, under map composition, forms a subgroup (the automorphism group on G) of S_n. We denote $\#H = m$.
3. $[g_i]$ is an *order equivalence class*, that is, g_i and g_j are order equivalent if and only if $g_i = g_j$. In this case, we shall also say that g_i and g_j are *relatively isomorphic*.

A major characteristic of order equivalence is given in the following theorem.

Theorem. g_i and g_j are relatively isomorphic if and only if there exists $h' \in H$ such that $h_j \cdot h_i^{-1} = h'$.

Proof: Assume $g_i = g_j$. Then,

$$h_i^{-1} \cdot g \cdot h_i = h_j^{-1} \cdot g \cdot h_j \Rightarrow g = h_i \cdot h_j^{-1} \cdot g \cdot h_j \cdot h_i^{-1}$$

But $h_j \cdot h_i^{-1} = h'$ for some $h' \in S_n$, and therefore, $g = h'^{-1} \cdot g \cdot h'$. It follows then that $h' \in H$.

The converse is straightforward.   Q.E.D.

Corollary

$$\#[g_j] = m, \qquad \text{for all } j = 1, \ldots, n!$$

Proof: By definition, $\#[g] = \#H = m$, and if we choose h_1 to be the identity bijection, we find $g = g_1 \Rightarrow \#[g_1] = m$. Thus it is sufficient to show $\#[g_j] = \#H$, for all $j = 2, \ldots, n!$, with $g_j \notin [g_1]$. Choose $g_j = g_{j'}$ and $g_j \notin [g_1]$. Then the above theorem guarantees that there exists $h' \in H$ such that

$$h_{j'} = h' \cdot h_j$$

In addition, for $h'' \in H$, $h'' \neq h'$, the theorem guarantees

$$h_{j''} = h'' \cdot h_j$$

and, therefore, $g_j = g_{j''}$. Iteration of this construction throughout H yields

$$[g_j] = \{g_j, g_{j'}, g_{j''}, \ldots, g_{j^{m-1'}}\}$$

and $\#[g_j] = \#H = m$. Since $[g_j]$ was chosen arbitrarily, the corollary follows.
                                                                          Q.E.D.

We are now in a position to prove the result stated in the introductory paragraphs of this section.

Proposition.   The number of order equivalence classes on a group $G := (S, g)$ is $n!/m$, where $n = \#S$ and m is the number of automorphisms of G.

Proof: From the corollary, $\#[g_i] = \#[g_j] = m$, for all $i, j = 1, \ldots, n!$. Hence the number of order equivalence classes is $\#S_n / \#[g_i] = n!/m$.   Q.E.D.

REMARK.   It is important to note, of course, that the above results are valid for any other distinct group $G' := (S, g')$ of order n. In this case, $\#[g_i'] = \#H' = m'$, where $H' \subset S_n$ is the automorphism group (under map composition) on G'. In general, for all groups of order n, the order equivalence classes are governed by the automorphism subgroups of S_n, and two distinct groups of order n cannot give rise to the same order equivalence classes. In fact, considering the structure of automorphism subgroups of S_n, it is relatively easy to show that, for all groups of order n, $n!/m \leqslant n!/(n-1)! = n$, thus giving a lower bound to the total number of order equivalence classes.

The primary aim of this section was to provide a fundamental exposition, both constructive and combinatorial, of order equivalence classes for finite groups of arbitrary order.

With respect to the *GGC*, given that* $\#H_V = 6$ and $\#H_{Z_4} = 2$, one may conclude, as had already been shown by brute force [5–8], that the numbers of order equivalence classes on V and Z_4 are 4 and 12, respectively. However, we are now in a position to extend the above treatment of order equivalence to a delineation of the combinatorial nature of partitions of C, the codon set. This work will ultimately result in a detailed specification of C partitions in a manner so as to assess, in closed form, the admissibility of f_i codes with respect to codon degeneracy symmetries.

5. Biological Contexts and *GGC* Universality

The mathematical treatment of the preceding sections was developed to formalize explicitly the symmetry patterns of the genetic code. We are now in a position to test the efficacy of this symmetry-searching ansatz through the analysis of reported alternative codings with those of their *SGC* counterparts.

The problem of alternative codings is as old as that of the genetic code itself. From the beginning it was clear that certain conditions could vitiate the *SGC* assignments [16]. For example, a large number of ambiguous codings result from in vitro studies in which a system is stressed by experimental conditions differing from those thought to exist in vivo. In addition, however, an increasing number of in vivo ambiguous assignments have also been established.

Regardless of their origins, the existence of alternative codings necessarily detracts from the concept of *SGC* universality. The *GGC* results in a reestablishment of code universality, albeit in a modified form. The necessity postulate permits an investigation of the symmetry characteristics of ambiguous codings within a format corresponding to that of the *SGC*, ambiguous assignments now being rationalized in terms of the relations R_1 and R_2. For example, codon pairs which satisfy the order-2 necessity postulate under at least one decomposition are considered to be doubly degenerate. This result facilitates unification of ambiguous codings with those derived from the f mapping. In fact, it is the prime intention of this development to view alternative codings on an equal basis with the *SGC*. The rationale behind this, of course, is the assumed reality of biological contexts under which code interpretation proceeds. For some contexts, however, certain codon assignments may remain the same as in the *SGC*, while others are modified.

*This may be easily seen by an inspection of V and Z_4: all automorphisms must leave the group relators invariant.

Table 5-5. Symmetries of Alternative Codings

Biological Context	Alternative Codings	SGC Equivalences	Decompositions	References
Amino acid starvation	$CAU \mapsto Gln$	CAG	D_6, D_7	[17]
	$CAC \mapsto Gln$	CAA	D_6, D_7	[17]
	$CGU \mapsto Cys$	UGU	$D_1 - D_5$	[18]
		UGC	D_0	[18]
	$CGC \mapsto Cys$	UGU	D_0	[18]
		UGC	$D_1 - D_3, D_6, D_7$	[18]
	$UGU \mapsto Trp$	UGG	D_6, D_7	[19]
Initiation	$GUG \mapsto f_{Met}$	AUG	$D_1 - D_5$	[20]
	$GUA \mapsto f_{Met}$	AUG	D_0	[20]
Missense suppression	$AGA \mapsto Gly$	GGU	D_6, D_7	[21–23]
		GGC	D_4, D_5	[21–23]
		GGA	$D_1 - D_3$	[21–23]
		GGG	D_0	[21–23]
	$UGU \mapsto Gly$	GGU	D_6, D_7	[24]
		GGC	D_0	[24]
		GGA	D_1, D_4, D_5	[24]
		GGG	D_2, D_3	[24]
	$GAU \mapsto Gly$	GGU	$D_1 - D_5$	[25]
		GGC	D_0	[25]
		GGA	D_6, D_7	[25]
	$GAC \mapsto Gly$	GGU	D_0	[25]
		GGC	$D_1 - D_3, D_6, D_7$	[25]
		GGA	D_4, D_5	[25]

Nonsense suppression		Codon		Ref.
	UAG→Ser	UCU	D_2, D_3	[26–28]
		UCC	D_1	[26–28]
		UCA	D_0, D_4, D_5	[26–28]
		UCG	D_6, D_7	[26–28]
		AGC	D_1-D_7	[26–28]
	UAG→Gln	CAA	D_0	[29, 30]
		CAG	D_1-D_3	[29, 30]
	UAG→Tyr	UAC	D_4, D_5	[28, 31, 32]
	UAG→Lys	AAG	D_4, D_5	[33]
		AAA	D_0	[33]
	UAG→Leu	UUA	D_0	[28, 33]
		CUU	D_2, D_3	[28, 33]
		CUC	D_1	[28, 33]
		CUA	D_0, D_4, D_5	[28, 33]
		CUG	D_6, D_7	[28, 33]
	UAG→Trp	UGG	D_1-D_5	[30]
	UGA→Trp	UGG	D_0-D_5	[34–36]
	UGA→Leu	UUG	D_0	[37]
		CUU	D_2, D_3	[37]
		CUC	D_1	[37]
		CUA	D_4, D_5	[37]
		CUG	D_0, D_6, D_7	[37]
	UGA→Ser	UCU	D_2, D_3	[37]
		UCC	D_1	[37]
		UCA	D_4, D_5	[37]
		UCG	D_0, D_6, D_7	[37]
		AGC	D_1-D_3	[37]
	UAA→Tyr	UAU	D_4, D_5	[28, 32, 38]
	UAA→Lys	AAG	D_0	[39, 40]
	UAA→Gln	CAA	D_1-D_7	[41]
		CAG	D_0	[41]

Table 5-5. (*Contd.*)

Biological Context	Alternative Codings	SGC Equivalences	Decompositions	References
	UAA↦Glu	GAG	D_0, D_4, D_5	[33]
	UAA↦Trp	UGG	D_0	[33]
	UAA↦Leu	UUA	D_4, D_5	[28]
		UUG	D_0	[28]
		CUU	D_1, D_6, D_7	[28]
		CUC	D_2, D_3	[28]
		CUG	D_0, D_4, D_5	[28]
	UAA↦Ser	UCU	D_1, D_6, D_7	[28]
		UCC	D_2, D_3	[28]
		UCG	D_0, D_4, D_5	[28]
		AGU	$D_1 - D_3, D_6, D_7$	[28]
Increased Mg^{+2} concentration.	UGU↦Tyr	UAU	$D_1 - D_7$	[42]
		UAC	D_0	[42]
	UGU↦Ser	UCC	D_0, D_6, D_7	[42]
		UCG	$D_1 - D_5$	[42]
		AGC	D_0, D_6, D_7	[42]
	UGU↦Arg	CGU	$D_1 - D_5$	[42]
		CGC	D_0	[42]
		CGA	D_6, D_7	[42]
		AGA	D_2, D_3	[42]
		AGG	D_1, D_4, D_5	[42]
Increased Mg^{+2} concentration, decreased temperature	UUU↦Leu	CUU	$D_1 - D_7$	[43, 44]
		CUC	D_0	[43, 44]
	UUU↦Ile	AUC	D_0	[44]
		AUA	D_2, D_3	[44]
	UUU↦Ser	AGU	$D_1, D_4 - D_7$	[44]
		AGC	D_0, D_2, D_3	[44]
		UCU	$D_1 - D_7$	[44]
		UCC	D_0	[44]

56

Condition	Mutation	Codon	D	Ref.
Presence of ethanol	$UUU \mapsto Tyr$	UAC	D_0	[44]
	$CCC \mapsto Leu$	UUA	D_6, D_7	[45]
		UUG	D_4, D_5	[45]
		CUU	D_0	[45]
		CUC	D_1-D_7	[45]
	$CCC \mapsto Thr$	ACU	D_0	[45]
		ACA	D_2, D_3	[45]
		ACG	D_1, D_4-D_7	[45]
	$UUU \mapsto Leu$	CUU	D_1-D_7	[45]
		CUC	D_0	[45]
	$UUU \mapsto Ile$	AUC	D_0	[45]
		AUA	D_2, D_3	[45]
Presence of antibiotics	$CGU \mapsto Cys$	UGU	D_1-D_5	[18]
		UGC	D_0	[18]
	$CGC \mapsto Cys$	UGU	D_0	[18]
		UGC	D_1-D_3, D_6, D_7	[18]
	$UUU \mapsto Leu$	CUU	D_1-D_7	[46, 47]
		CUC	D_0	[46, 47]
	$UUU \mapsto Ile$	AUC	D_0	[46, 47]
		AUA	D_2, D_3	[46, 47]
	$UUU \mapsto Ser$	AGU	D_1, D_4-D_7	[46, 47]
		AGC	D_0, D_2, D_3	[46, 47]
		UCU	D_1-D_7	[46, 47]
		UCC	D_0	[46, 47]
	$UCU \mapsto Phe$	UUU	D_1-D_7	[48]
		UUC	D_0	[48]
	$UCU \mapsto Pro$	CCU	D_1-D_3	[48]
		CCC	D_0	[48]
		CCA	D_6, D_7	[48]
		CCG	D_4, D_5	[48]

57

Table 5-5. (*Contd.*)

Biological Context	Alternative Codings	SGC Equivalences	Decompositions	References
	UCU↦His	CAU	D_4, D_5	[48]
		CAC	D_0, D_6, D_7	[48]
	UCU↦Arg	CGU	D_6, D_7	[48]
		CGC	D_0, D_4, D_5	[48]
		CGA	D_1	[48]
		CGG	D_2, D_3	[48]
		AGA	D_4, D_5	[48]
		AGG	D_6, D_7	[48]
	UCU↦Ile	AUC	D_0, D_4, D_5	[48]
		AUA	D_1, D_6, D_7	[48]
	UCU↦Thr	ACU	D_4, D_5	[48]
		ACC	D_0, D_6, D_7	[48]
		ACA	D_2, D_3	[48]
		ACG	D_1	[48]
	CUC↦Phe	UUU	D_0	[48]
		UUC	D_1-D_3	[48]
	CUC↦Pro	CCU	D_0	[48]
		CCC	D_1-D_7	[48]
	CUC↦His	CAU	D_0, D_6, D_7	[48]
	CUC↦Arg	CGU	D_0, D_4, D_5	[48]
		CGA	D_2, D_3	[48]
		CGG	D_1, D_6, D_7	[48]
		AGA	D_6, D_7	[48]
		AGG	D_4, D_5	[48]
	CUC↦Ile	AUU	D_0, D_4, D_5	[48]
		AUC	D_6, D_7	[48]
		AUA	D_2, D_3	[48]

Mutation	Codon	D	Ref
CUC→Thr	ACU	D_0, D_6, D_7	[48]
	ACA	D_1, D_4, D_5	[48]
	ACG	D_2, D_3	[48]
UGU→Arg	CGU	D_1-D_5	[48]
	CGC	D_0	[48]
	CGA	D_6, D_7	[48]
	AGA	D_2, D_3	[48]
	AGG	D_1, D_4, D_5	[48]
UGU→Ser	UCC	D_0, D_6, D_7	[48]
	UCG	D_1-D_5	[48]
	AGC	D_0, D_6, D_7	[48]
UGU→Gln	CAA	D_4, D_5	[48]
	CAG	D_6, D_7	[48]
UGU→Leu	UUA	D_1-D_7	[48]
	CUC	D_0, D_6, D_7	[48]
	CUG	D_1-D_5	[48]
UGU→Tyr	UAU	D_1-D_7	[48]
	UAC	D_0	[48]
UGU→Pro	CCU	D_6, D_7	[48]
	CCC	D_0, D_4, D_5	[48]
	CCA	D_1-D_3	[48]
GUG→Arg	CGA	D_0	[48]
	CGG	D_1-D_7	[48]
	AGA	D_0, D_6, D_7	[48]
GUG→Ser	UCU	D_1, D_4, D_5	[48]
	UCC	D_2, D_3	[48]
	UCA	D_0, D_6, D_7	[48]
	AGU	D_1-D_5	[48]
GUG→Gln	CAA	D_0-D_3	[48]
GUG→Leu	UUA	D_0	[48]
	UUG	D_6, D_7	[48]
	CUU	D_1, D_4, D_5	[48]
	CUC	D_2, D_3	[48]
	CUA	D_0, D_6, D_7	[48]

Table 5-5. (*Contd.*)

Biological Context	Alternative Codings	SGC Equivalences	Decompositions	References
	GUG↦Tyr	UAC	D_6, D_7	[48]
	GUG↦Pro	CCU	D_2, D_3	[48]
		CCC	D_1, D_6, D_7	[48]
		CCA	D_0, D_4, D_5	[48]
	CAC↦Arg	CGU	D_0	[48]
		CGC	D_1-D_7	[48]
		AGA	D_1, D_4, D_5	[48]
		AGG	D_2, D_3	[48]
	CAC↦Gln	CAA	D_6, D_7	[48]
	CAC↦Tyr	UAU	D_0	[48]
		UAC	D_1-D_5	[48]
	ACA↦Arg	CGU	D_6, D_7	[48]
		CGA	D_1-D_5	[48]
		CGG	D_0	[48]
		AGG	D_0, D_6, D_7	[48]
	ACA↦Gln	CAG	D_0-D_5	[48]
	AGA↦Gln	CAG	D_0, D_6, D_7	[48]
	AGA↦His	CAU	D_2, D_3	[48]
		CAC	D_1, D_4, D_5	[48]
	AGA↦Ile	AUU	D_1, D_6, D_7	[48]
		AUC	D_2, D_3	[48]
	AGA↦Pro	CCU	D_6, D_7	[48]
		CCA	D_1, D_4, D_5	[48]
		CCG	D_0, D_2, D_3	[48]
	AGA↦Thr	ACU	D_2, D_3	[48]
		ACC	D_1, D_4, D_5	[48]
		ACG	D_0, D_6, D_7	[48]

60

AGA→Val	GUU	D_2, D_3	[48]
	GUC	D_1	[48]
	GUA	D_4, D_5	[48]
	GUG	D_0, D_6, D_7	[48]
GAG→Gln	CAA	D_0, D_6, D_7	[48]
	CAG	D_4, D_5	[48]
GAG→His	CAU	D_1	[48]
	CAC	D_2, D_3	[48]
GAG→Ile	AUU	D_2, D_3	[48]
	AUC	D_1	[48]
	AUA	D_0, D_4, D_5	[48]
GAG→Pro	CCU	D_4, D_5	[48]
	CCA	D_0, D_2, D_3	[48]
	CCG	D_1, D_6, D_7	[48]
GAG→Thr	ACU	D_1	[48]
	ACC	D_2, D_3	[48]
	ACA	D_0, D_6, D_7	[48]
	ACG	D_4, D_5	[48]
GAG→Val	GUU	D_1, D_4, D_5	[48]
	GUC	D_2, D_3	[48]
	GUA	D_0, D_6, D_7	[48]
CUA→Asp	GAU	D_4, D_5	[49]
	GAC	D_6, D_7	[49]
CUA→Gln	CAG	D_0, D_6, D_7	[49]
CUA→Lys	AAA	$D_1, D_4 - D_7$	[50]
	AAG	D_0, D_2, D_3	[50]
UCA→His	CAC	$D_1 - D_5$	[49]
CCA→His	CAU	$D_1 - D_5$	[49]

In vitro
system, context
unknown

61

Table 5-5. (*Contd.*)

Biological Context	Alternative Codings	SGC Equivalences	Decompositions	References
	CCG↦Arg	CGU	$D_1 - D_3, D_6, D_7$	[49]
		CGA	D_0	[49]
		CGG	D_4, D_5	[49]
		AGA	D_0, D_2, D_3	[49]
		AGG	D_1, D_6, D_7	[49]
	GCG↦Arg	CGC	D_4, D_5	[49]
		CGA	$D_0 - D_3, D_6, D_7$	[49]
		AGA	D_0, D_4, D_5	[49]
	UAA↦Lys	AAG	D_0	[49]
	UAG↦Asp	GAC	$D_1 - D_3$	[49]
	CAU↦Tyr	UAU	$D_1 - D_3, D_6, D_7$	[49]
		UAC	D_0	[49]
	CAC↦Thr	ACU	$D_0 - D_5$	[49]
		ACA	D_6, D_7	[49]
	CAG↦Asp	GAU	$D_1 - D_3$	[49]
	AAC↦Thr	ACU	D_0	[49]
		ACC	D_6, D_7	[49]
		ACG	$D_1 - D_5$	[49]
	GAG↦Asp	GAU	D_6, D_7	[49]
		GAC	D_4, D_5	[49]
	UGU↦Trp	UGG	D_6, D_7	[49]
	UGU↦Val	GUC	$D_0 - D_5$	[49]
		GUG	D_6, D_7	[49]
	UGA↦Arg	CGU	D_6, D_7	[49]
		CGC	D_4, D_5	[49]
		CGA	$D_1 - D_3$	[49]
		CGG	D_0	[49]
		AGA	D_4, D_5	[49]
		AGG	D_0, D_6, D_7	[49]

UGA→Asn	AAU	D_1-D_3, D_6, D_7	[49]
UGA→Asp	GAC	D_1-D_3	[49]
UGA→Cys	UGC	D_6, D_7	[49]
UGA→Gln	CAG	D_0-D_3	[49]
UGA→Glu	GAA	D_4, D_5	[49]
	GAG	D_0, D_6, D_7	[49]
UGG→Gly	GGC	D_1-D_7	[49]
	GGA	D_0	[49]
CGU→Cys	UGU	D_1-D_5	[49]
	UGC	D_0	[49]
CGU→Val	GUU	D_1-D_5	[49]
	GUC	D_0	[49]
	GUA	D_6, D_7	[49]
CGG→Gly	GGU	D_1-D_7	[49]
	GGA	D_0	[49]
AGU→Cys	UGC	D_0, D_6, D_7	[49]
AGC→Ala	GCU	D_0, D_4, D_5	[49]
	GCG	D_1-D_3, D_6, D_7	[49]
AGC→Cys	UGU	D_0, D_6, D_7	[49]
	UGC	D_4, D_5	[49]
AGG→Gly	GGA	D_0	[49]
	GGG	D_1-D_7	[49]
AGG→Phe	UUC	D_6, D_7	[49]
GGC→Arg	CGU	D_0	[49]
	CGC	D_4, D_5	[49]
	CGA	D_1-D_3, D_6, D_7	[49]
	AGA	D_4, D_5	[49]
GGA→Glu	GAA	D_1-D_3	[49]
	GAG	D_0	[49]
GGA→Asp	GAU	D_6, D_7	[49]
	GAC	D_4, D_5	[49]

Table 5-6. Symmetries of Mitochondrial Alternative Codings

Alternative Codings	Coding Equivalences[a]	Partitions	Species	References
UGA↦Trp	UGG	D_0–D_5	Human, bovine, mouse, Drosophila, Saccharomyces, Aspergillus, Neurospora	[51–64]
AUA↦Met	AUG	D_0–D_3, D_6, D_7	Human, bovine, mouse, Drosophila, Saccharomyces	[51, 54, 58–60, 62–64]
AGA↦Ser	UCU	D_4–D_7	Drosophila	[64]
	UCC	D_6, D_7		[64]
	UCA	D_2, D_3		[64]
	UCG	D_0, D_1		[64]
	AGU	D_4, D_5		[64]
	AGC	D_6, D_7		[64]
CUA↦Thr	ACC	D_1–D_5	Saccharomyces	[57, 63, 65–67]
	ACG	D_0, D_6, D_7		[57, 63, 65–67]
CUG↦Thr	ACU	D_1–D_3	Saccharomyces	[57, 63]
	ACA	D_0, D_6, D_7		[57, 63]
	ACG	D_4, D_5		[57, 63]
CUC↦Thr	ACU	D_0, D_6, D_7	Saccharomyces	[57, 63]
	ACA	D_1, D_4, D_5		[57, 63]
	ACG	D_2, D_3		[57, 63]

CUU↦Thr	ACU	D_4, D_5	Saccharomyces	[57, 63]
	ACC	D_0, D_6, D_7		[57, 63]
	ACA	D_2, D_3		[57, 63]
	ACG	D_1		[57, 63]
AGA↦TC	UGA	D_4, D_5	Human, bovine	[59, 62]
	UAG	D_0, D_4, D_5		[59, 62]
AGG↦TC	UGA	D_0, D_6, D_7	Human, bovine	[59, 62]
	UAA	D_0, D_4, D_5		[59, 62]
	UAG	D_6, D_7		[59; 62]
AUA↦initiation	AUG	D_0-D_3, D_6, D_7	Human, bovine, mouse, Drosophila	[59, 60, 62, 64]
AUU↦initiation	AUU	D_4, D_5	Human, mouse, Drosophila	[59, 60, 62, 64]
	AUC	D_0-D_3, D_6, D_7		[59, 60, 64]
AUC↦initiation	AUA	D_4, D_5	Mouse	[59, 60, 64]
	AUG	D_4, D_5		[60]
	AUU	D_0-D_3, D_6, D_7		[60]

aCoding equivalences used are those of the SGC except for AUA, AUU, AUC↦initiation, where the coding equivalences used are those of the remaining codons of the AUN quartet.

65

Nonetheless, this modification must satisfy the necessity postulate if codon degeneracies are to result.

An extensive listing of presently known alternative codings is presented, and analyzed within the GGC, in Tables 5-5 and 5-6. That the GGC analysis is successful in every case provides strong support for the existence of multiple biological coding contexts. Such evidence, in fact, functions as a critical test of the GGC. (These data are not intended as a thoroughly exhaustive listing of ambiguous codings but, rather, as an illustrative assemblage of the GGC operative under different biological contexts.) An alternative coding assignment, even if analyzed with respect only to its SGC equivalence (i.e., equivalence under f mapping), forms, conjunctively with the SGC counterparts, one or more codon pairs which are degenerate under certain GGC decompositions. A close examination of Tables 5-5 and 5-6 shows that certain codon pairs satisfy the order-2 necessity postulate for D_4, D_5, D_6 or D_7 without satisfying it for D_1-D_3. This effectively validates the viewpoint that D_4-D_7 are appropriate for symmetries of codon degeneracies under biological contexts differing from the SGC context.

The evidence for in vitro alternative codings is easily obtained from studies involving significantly stressed experimental systems. Various environmental parameters, including changes in temperature, ion concentration, and the presence of antibiotics or organic solvents (see Table 5-5), have been identified as distinct biological contexts promoting recognized ambiguities. Until quite recently, however, in vivo alternative assignments were restricted to missense and nonsense suppression studies (see Table 5-5). In nonsense suppression, a means is provided for competing with chain termination by inserting an amino acid at an SGC termination codon (UAA, UAG, UGA) site; missense suppression involves an alternative reading of a codon specifying a particular amino acid. Nonsense suppression, in which UGA codes for Trp [34–36], is of particular interest. UGA and UGG, the SGC tryptophan codon, constitute a degenerate codon pair which satisfies the necessity postulate under D_0-D_5. In addition, such a Trp coding partially removes the SGC symmetry deviation generated by odd-order degenerate codons, as discussed previously.

The increasing number of reported in vivo ambiguities involving codons other than those derived from suppression is of critical importance in identifying additional contexts for protein synthesis. Bacterial or mammalian cells starved for a particular amino acid appear to compensate by utilizing a different amino acid. These studies [17, 18] involved either electrophoretic analysis [17] or radioisotope techniques [18] to determine the effects of amino acid starvation and, therefore, do not provide direct evidence for amino acid substitution. Such documentation must come from the sequencing of proteins synthesized during starvation. However, the initial evidence (see Table 5-5) strongly suggests that amino acid starvation may be a distinct biological context for translation.

Mitochondrial DNA (mtDNA) sequence analyses, when combined with independent determinations of the amino acid sequences of the resultant proteins, provide the most recent and, thus far, the most extensive roster of in vivo alternative codon assignments. Each of these alternative codings may be successfully analyzed according to the symmetry of the *GGC* (see Table 5-6). The data of Table 5-6 serve to complement, as well as to extend, the presently known cadre of alternative codings derived from diverse biological contexts operative during cytoplasmic protein synthesis (see Table 5-5). For example, in addition to the mitochondrial assignments of Table 5-6, UGA is known to code for the amino acid tryptophan during nonsense suppression [34–36], while CUA codes for threonine in the presence of antibiotics [48]. However, the remaining alternative codings listed in Table 5-6 are unique to mitochondrial code interpretation.

In addition to further illustrating the prevalence of alternative codings, and the efficacy of the symmetry analysis of these codings via the partitions of the *GGC*, Table 5-6 also indicates interspecific variations in mitochondrial code interpretation. All of the mitochondrial systems surveyed employ UGA as a tryptophan codon, as opposed to its *SGC* assignment as a termination codon. However, certain alternative codings occur principally in the lower eukaryotes, while others appear restricted to higher organisms, including humans. For example, *Drosophila*, as well as certain mammalian systems, exploit novel termination and/or initiation codons, while simple fungal organisms seem to be confined to the *SGC* equivalences for these functions. Similarly, assignment of the CUN ($N =$ A, U, C, G) quartet to threonine occurs in the yeast *Saccharomyces*, while in higher organisms CUN assumes its *SGC* leucine designation. Nevertheless, even among closely related organisms, such as the fungi, divergent codon assignments occur. Yeast mitochondria, for example, unlike those from *Neurospora* and *Aspergillus*, employ AUA as a methionine codon. In addition, the assignment of CUN to threonine, within the fungi, is also apparently confined to yeast.

Such interspecific variations in mitochondrial code interpretation underscore the importance of the biological context as an integral facet of the genetic code. Consequently, mitochondrial protein synthesis does *not* represent a distinct biological context but, rather, must constitute a class of disparate biological contexts, each of which serves to delimit a species-specific mitochondrial genetic code.

The mitochondrial alternative codings reported thus far also provide critical input into the question of the basic symmetry of the standard genetic code itself. The analysis provided by Section 3.A identified the primitive symmetry of the genetic code as residing in the order-2 degenerate codons and explained odd-order degenerate codons of the *SGC* as examples of slight deviations from the observed symmetry. The mitochondrial codings UGA $\mapsto$ trp and AUA $\mapsto$ met are of particular relevance to this argument: when both of these alternative codings are exhibited simultaneously in a particular mitochondrial genetic code (as in the case of all systems surveyed in Table 5-6, with the exception of *Aspergillus* and *Neurospora*), the broken symmetry of the *SGC*, generated by the odd-order degenerate codons, is removed.

6. Conclusions

The discovery of the *SGC* (i.e., that code resulting from the f mapping) is a major accomplishment of modern molecular genetics. However, the existence of alternative codon assignments has acted as an impass to a consistent interpretation of the code. In view of the ever-increasing number and types of such "ambiguous" codings, the standard genetic code can no longer be considered universal. With this in mind, we are faced with the options of either (1) handling each ambiguity, old or new, as an individual, exceptional event, the result of an unhappy accident during protein synthesis, or (2) reinvestigating the symmetry characteristics of the genetic code in order to define some more extended structure which can accommodate *SGC* breakdown and provide evidence for the unification of standard genetic code assignments with ambiguous codon assignments. The purpose of the present work has been to examine the latter possibility.

Although we certainly have not completely resolved this problem, we feel we have provided substantial new insight in terms of the following conclusions.

1. The basic symmetry of the *SGC* has been rephrased in terms of the sets F_1, F_2 [Eqs. (5-9a, b)]. Such rephrasing exposes the primitive symmetry as residing in the order-2 degenerate codons. The symmetry of the higher even-order degeneracies results from invoking the primitive symmetry by conjunctive and disjunctive application of Eqs. (5-9a, b).

2. The group-theoretic processing developed in this work allows a consistent generalization of the symmetry characteristics of the *SGC* with the symmetry exhibited by alternative codon assignments. What results are the necessary (but not sufficient) conditions for codon degeneracy [Eqs. (5-10a, b)].

3. The generalized approach to the genetic code sketched by Gatlin [12] implies that the total code be envisioned initially as $C \times A$. A biological context then serves to select a particular subset of $C \times A$ (e.g., in the case

of the standard code, the subset chosen is that defined by the f mapping). Such a generalized code appears to be logically necessary if ambiguous codon assignments are to be considered as an integral part of the genetic apparatus. It then follows that the code must be considered a priori as a much more complicated structure than was previously thought. What has been demonstrated here is the possible existence of a constraint on $C \times A$ that must necessarily serve to restrict the biologically meaningful subsets of $C \times A$. Thus while the genetic code may very well be more complex than the standard view implies, it does exhibit well-defined regularities as vested in the partitions of the GGC (see Table 5-4).

4. The supposition of biological contexts is central to a coherent view of the GGC. Indeed, the very existence of biological contexts becomes operationally mandated by SGC breakdown if mapping contexts are to be retained at all. A context restricts the GGC to one of many specific codes by confining the GGC relation to a fixed mapping. With the introduction of context structure, then, the question of possible universality of the genetic code must be rephrased in terms of the possible universality of a specific biological context. It is important to point out that coding contexts, as operationally phrased above, have not been well studied. (Indeed, strict adherence to the SGC prohibits even the *conceptualization* of a context.) Such studies are manifestly needed if the coding problem is ever to receive a comprehensive understanding.

The experimental data which form the basis for the symmetry analyses of Tables 5-5 and 5-6 are derived, in large part, from reports of complementary DNA *and* amino acid analyses. Ever since the advent, in 1977, of rapid DNA-sequencing techniques [68], an increasing number of DNA sequences from a broad spectrum of organisms and from disparate regions of the genome have been reported. Unfortunately, in many cases the amino acid sequence of the resultant protein is constructed directly from the DNA base sequence under the assumption of the SGC. In light of the accumulating alternative codings from a wide variety of sources, it is inevitable that many of the reported amino acid sequences will prove incorrect when subjected to independent verification. In order to avoid this pitfall, reports of protein sequence data derived from DNA analyses should include a caveat, namely, that the amino acid sequence represents a probable assignment (i.e., that of the SGC) which may be modified by the introduction of alternative codings resulting from the imposition of alternative biological contexts.

In particular, consideration of mitochondrial alternative codings clearly indicates that mitochondrial code interpretation constitutes a composite of diverse biological contexts and hence a class of distinct, species-specific mitochondrial genetic codes. Consequently, in addition to providing support for the conjecture of independent evolution of mitochondrial and cytoplasmic protein synthesis, these data also generate significant insight into the question of mitochondrial evolution itself. The usual explanation of mitochondrial origins invokes some form of endosymbiosis, involving colonization of a primitive eukaryotic cell by a respiring, bacteria like organism [69]. The existence of multiple mitochondrial genetic codes, however, indicates that this event (1) may have occurred, multiply and independently, with several different colonizing organisms possessed of distinct genetic codes, or (2) may have transpired in a more concerted fashion (i.e., involving a single endosymbiont) followed by divergence of the genetic code as a result of selective pressures experienced by the captive cellular organelle in its new host environment. In any case, based upon evidence from current mitochondrial genetic codes, the endosymbiont may have been no more closely related to present-day bacteria than to higher organisms!

In summary, then, while the genetic code is surely more complex than the standard view implies, it nevertheless exhibits the well-defined structural regularities vested in the partitions of the *GGC*. By focusing on the symmetry constraints of the problem, therefore, the *GGC* supplies an effective ansatz for the systematic analysis of alternative codings. As a result, the *GGC* provides a reunification of alternative codon:amino acid assignments with those of the *SGC*. An essential aspect of the *GGC* is the admission of multiple genetic codes which are uniquely interpreted via distinct biological contexts. Consequently, a single genetic message interpreted under multiple biological contexts will result in transcripts of multiple coding composition (i.e., different proteins). Therefore the imposition of multiple coding contexts—and hence multiple genetic codes—increases the potential information content of the coding process. A complete explication of the formal operative nature of biological contexts within genetic coding theory, as well as a delineation of the full consequences of contextual increases in the information content of the genetic code, requires an explicit group-theoretic description of all possible alternative codes which satisfy the generalized codon-degeneracy symmetry.

References

1. R. V. Eck, *Science*, **140**, 477 (1963).
2. C. R. Woese, *Proc. Natl. Acad. Sci. USA*, **54**, 71 (1965).
3. F. H. C. Crick, *J. Mol. Biol.*, **38**, 367 (1968).
4. T. H. Jukes, in *Comprehensive Biochemistry*, vol. 24, M. Florkin, A. Neuberger, and L. L. M. van Deenen, Eds., Elsevier, Amsterdam, 1977, p. 235.

5. G. L. Findley and S. P. McGlynn, *Int. J. Quantum Chem., Quantum Biol. Symp.*, **6**, 313 (1979).

6. G. L. Findley and S. P. McGlynn, *Int. J. Quantum Chem., Quantum Biol. Symp.*, **7**, 277 (1980).

7. A. M. Findley, G. L. Findley, and S. P. McGlynn, *J. Theor. Biol.*, **97**, 299 (1982).

8. G. L. Findley, A. M. Findley, and S. P. McGlynn, *Proc. Natl. Acad. Sci. USA*, **79**, 7061 (1982).

9. A. M. Findley and G. L. Findley, *Int. J. Quantum Chem., Quantum Biol. Symp.*, **9**, 59 (1982).

10. A. M. Findley and G. L. Findley, *Ann. N. Y. Acad. Sci.*, **435**, 537 (1984).

11. A. M. Findley and G. L. Findley, *Int. J. Quantum Chem., Quantum Biol. Symp.*, **11**, 109 (1984).

12. L. L. Gatlin, *Information Theory and the Living System*, Columbia University Press, New York, 1972.

13. F. H. C. Crick, *J. Mol. Biol.*, **19**, 548 (1966).

14. W. Magnus, A. Karrass, and D. Solitar, *Combinatorial Group Theory*, 2nd ed., Dover, New York, 1976.

15. F. H. C. Crick, S. Brenner, A. Klug, and C. Preczenik, *Origins Life*, **7**, 389 (1976).

16. M. Ỹcas, *The Biological Code*, North-Holland, Amsterdam, 1969.

17. J. Parker, J. W. Pollard, J. D. Friesen, and C. P. Stanners, *Proc. Natl. Acad. Sci. USA*, **75**, 1091 (1978).

18. P. Edelmann and J. Gallant, *Cell*, **10**, 131 (1977).

19. R. H. Buckingham and C. G. Kurland, *Proc. Natl. Acad. Sci. USA*, **74**, 5496 (1977).

20. H. P. Ghosh, D. Söll, and H. G. Khorana, *J. Mol. Biol.*, **25**, 275 (1967).

21. J. Carbon, C. Squires, and C. W. Hill, *Cold Spring Habor Symp. Quantum Biol.*, **34**, 505 (1969).

22. C. W. Hill, J. Foulds, L. Söll, and P. Berg, *J. Mol. Biol.*, **39**, 563 (1969).

23. C. W. Hill, C. Squires, and J. Carbon, *J. Mol. Biol.*, **52**, 557 (1970).

24. N. K. Gupta and H. G. Khorana, *Proc. Natl. Acad. Sci. USA*, **56**, 772 (1966).

25. H. Berger and C. Yanofsky, *Science*, **156**, 394 (1967).

26. D. L. Engelhardt, R. E. Webster, R. C. Wilhelm, and N. D. Zinder, *Proc. Natl. Acad. Sci. USA*, **54**, 1791 (1965).

27. M. R. Capecchi and G. N. Gussin, *Science*, **149**, 417 (1965).

28. P. W. Piper, in *tRNA: Biological Aspects*, D. Söll, J. N. Abelson, and P. R. Schimmel, Eds., Cold Spring Harbor Laboratory, Cold Spring Harbor, NY, 1980, p. 379.

29. L. Söll and P. Berg, *Nature*, **223**, 1340 (1969).

30. L. Söll and P. Berg, *Proc. Natl. Acad. Sci. USA*, **63**, 392 (1969).

31. H. M. Goodman, J. Abelson, A. Landy, S. Brenner, and J. D. Smith, *Nature*, **217**, 1019 (1968).

32. M. G. Weigert, E. Lanka, and A. Garen, *J. Mol. Biol.*, **23**, 401 (1967).

33. H. Ozeki, H. Inokuchi, F. Yamao, M. Kodaira, H. Sakano, T. Ikemura, and Y. Shimura, in *tRNA: Biological Aspects*, D. Söll, J. N. Abelson, and P. R. Schimmel, Eds., Cold Spring Harbor Laboratory, Cold Spring Harbor, NY, 1980, p. 341.

34. T.-S. Chan and A. Garen, *J. Mol. Biol.*, **49**, 231 (1970).

35. D. Hirsh, *J. Mol. Biol.*, **58**, 439 (1971).

36. D. Hirsh and L. Gold, *J. Mol. Biol.*, **58**, 459 (1971).

37. J. Kohli, F. Altruda, T. Kwong, A. Rafalski, R. Wetzel, D. Söll, G. Wahl, and U. Leupold, in *tRNA: Biological Aspects*, D. Söll, J. N. Abelson, and P. R. Schimmel, Eds., Cold Spring Harbor Laboratory, Cold Spring Harbor, NY, 1980, p. 407.

38. H. M. Goodman, M. V. Olson, and B. D. Hall, *Proc. Natl. Acad. Sci. USA*, **74**, 5453 (1977).

39. A. Garen, *Science*, **160**, 149 (1968).

40. S. Kaplan, *J. Bacteriol.*, **105**, 984 (1971).

41. H. Inokuchi and H. Dzeki, *Jpn. J. Genet. (Abstr.)*, **51**, 412 (1976).

42. S. Nishimura, F. Harada, and M. Hirabayashi, *J. Mol. Biol.*, **40**, 173 (1969).

43. S. M. Friedman and I. B. Weinstein, *Biochim. Biophys. Acta*, **114**, 593 (1966).

44. W. Szer and S. Ochoa, *J. Mol. Biol.*, **8**, 823 (1964).

45. A. G. So and E. W. Davies, *Biochemistry*, **3**, 1165 (1964).

46. J. Davis, W. Gilbert, and L. Gorini, *Proc. Natl. Acad. Sci. USA*, **51**, 883 (1964).

47. L. Gorini, in *Ribosomes*, M. Nomura, A. Tissieres, and P. Lengyl, Eds., Cold Spring Harbor Laboratory, Cold Spring Harbor, NY, 1974, p. 791.

48. J. Davies, D. S. Jones, and H. G. Khorana, *J. Mol. Biol.*, **18**, 48 (1966).

49. D. Söll, E. Ohtsuka, D. S. Jones, R. Lohrmann, H. Hayatsu, S. Nishimura, and H. G. Khorana, *Proc. Natl. Acad. Sci. USA*, **54**, 1378 (1965).

50. R. Brimacombe, J. Trupin, M. Nirenberg, P. Leder, M. Bernfield, and T. Jaouni, *Proc. Natl. Acad. Sci. USA*, **54**, 954 (1964).

51. B. G. Barrell, A. T. Bankier, and J. Drouin, *Nature*, **282**, 189 (1979).

52. G. Coruzzi and A. Tzagoloff, *J. Biol. Chem.*, **254**, 9324 (1979).

53. G. Macino, G. Coruzzi, F. G. Nobrega, M. Li, and A. Tzagoloff, *Proc. Natl. Acad. Sci, USA*, **76**, 3784 (1979).

54. B. G. Barrell, S. Anderson, A. T. Bankier, M. H. L. de Bruijn, E. F. Chen, P.H. Schreier, A. J. H. Smith, R. Staden, and I. G. Young, *Proc. Natl. Acad. Sci. USA*, **77**, 3164 (1980).

55. J. E. Heckman, J. Sarnoff, B. Alzner-DeWeerd, S. Yin, and U. L. Raj Bhandary, *Proc. Natl. Acad. Sci. USA*, **77**, 3159 (1980).

56. N. C. Martin, H. D. Pham, K. Underbrink-Lyon, D. L. Miller, and J. E. Donelson, *Nature*, **285**, 579 (1980).

57. B. E. Thalenfeld and A. Tzagoloff, *J. Biol. Chem.*, **255**, 6173 (1980).

58. I. G. Young and S. Anderson, *Gene*, **12**, 257 (1980).

59. S. Anderson, A. T. Bankier, B. G. Barrell, M. H. L. de Bruijn, A. R. Coulson, J.

Drouin, I. C. Eperon, D. P. Nierlich, B. A. Roe, F. Sanger, P. H. Schreier, A. J. Smith, R. Staden, and I. G. Young, *Nature*, **290**, 457 (1981).

60. M. J. Bibb, R. A. Van Etten, C. T. Wright, M. W. Walberg, and D. A. Clayton, *Cell*, **26**, 167 (1981).

61. R. B. Waring, R. W. Davies, S. Lee, E. Grisi, M. McPhail Berks, and C. Scazzocchio, *Cell*, **27**, 4 (1981).

62. S. Anderson, M. H. L. de Bruijn, A. R. Coulson, I. C. Eperon, F. Sanger, and I. G. Young, *J. Mol. Biol.*, **156**, 683 (1982).

63. M. E. S. Hudspeth, W. M. Ainley, D. S. Schumard, R. A. Butlow, and L. I. Grossman, *Cell*, **30**, 617 (1982).

64. M. H. L. de Bruijn, *Nature*, **304**, 234 (1983).

65. W. Sebald and E. Wachter, in *Energy Conservation in Biological Membranes*, G. Schafer and M. Klingenberg, Eds., Springer, New York, 1978, p. 228.

66. M. Li and A. Tzagoloff, *Cell*, **18**, 47 (1979).

67. G. Macino and A. Tzagoloff, *J. Biol. Chem.*, **254**, 4617 (1979).

68. F. Sanger, S. Nicklen, and A. R. Coulson, *Proc. Natl. Acad. Sci. USA*, **74**, 5463 (1977).

69. See, for example, J. F. Fredrick, Ed., *Origins and Evolution of Eukaryotic Intracellular Organelles*, *Ann. N. Y. Acad. Sci.*, **361** (1981).

PART 3 STATICS

6 Linear Spaces

1. Multistructure Objects

We now extend our previous discussion of sets and their structures (Chapter 3) to consider objects possessing two structure maps. Again, we employ the functions $f: S \times S \rightarrow S$ and $g: S \times S \rightarrow S$, where (S, f) satisfies the conditions for an abelian group. In addition, we define a new set $S' = S - Idf$, which consists of all the elements of S except for the identity. If (S', g) is at least a semigroup, we may impose the following compatibility axiom:

A5. $\forall x, y, z \in S, \qquad g(x, f(y, z)) = f(g(x, y), g(x, z))$

$$g(f(y, z), x) = f(g(y, x), g(z, x))$$

(A5) describes the distributivity of g over f (e.g., multiplication over addition). Whenever two structure maps interact, the condition of distributivity must be fulfilled. Given this condition, we may use multistructure maps to compose more complex object types.

A *ring* is a set S having two associated maps f and g such that (S, f) is an abelian group, (S', g) is a semigroup, and (S, f, g) satisfies (A5). The set of integers taken under arithmetic addition $(\mathbb{Z}, +)$ and multiplication $(\mathbb{Z}', \cdot)$ constitutes the ring $(\mathbb{Z}, +, \cdot)$. More precisely, since $(\mathbb{Z}', \cdot)$ is an abelian semigroup, $(\mathbb{Z}, +, \cdot)$ is designated a commutative ring.

A *field* is a set S having two maps f and g such that both (S, f) and (S', g) are abelian groups and (S, f, g) satisfies (A5). The sets of rational, real, and complex numbers, under arithmetic addition and multiplication, form the fields $(\mathbb{Q}, +, \cdot)$, $(\mathbb{R}, +, \cdot)$, and $(\mathbb{C}, +, \cdot)$, respectively.

It is also possible to construct a set consisting of the Cartesian product of multistructure objects having the form

$$\Pi_n = (S, f, g)_1 \times (S, f, g)_2 \times \cdots \times (S, f, g)_n$$

Let $\alpha, \beta \in (S, f, g)$ and $x, y \in \Pi_n$; and let $x = (\alpha_1, \alpha_2, \ldots, \alpha_n)$ and $y = (\beta_1, \beta_2, \ldots, \beta_n)$. We define the map $\Gamma: \Pi_n \times \Pi_n \rightarrow \Pi_n$, where Γ is interpreted as follows.

$\forall x, y \in \Pi_n, z \in \Pi_n$, such that

$$\Gamma(x, y) = (f(\alpha_1, \beta_1), f(\alpha_2, \beta_2), \ldots, f(\alpha_n, \beta_n))$$
$$= (\gamma_1, \gamma_2, \ldots, \gamma_n) = z \in \Pi_n$$

Clearly, Γ relates elements of Π_n to elements of (S, f, g). (Π_n, Γ) is denoted a Cartesian product object and it may be utilized to organize additional object types. If (Π_n, Γ) is an abelian group and (S, f, g) is a ring, the *n-module* M^n results. When (Π_n, Γ) is an abelian group and (S, f, g) is a field, an *n*-dimensional *linear space* V^n is formed.

2. Vector Spaces

We now consider in further detail the properties of linear spaces. Conventionally, one uses the notation $f \leftrightarrow +$, $g \leftrightarrow \cdot$ and $\Gamma \leftrightarrow +$ (cf. Chapter 3). By definition, a vector (or linear) space over the field $(S, +, \cdot)$ is an additive abelian group $(\Pi_n, +)$ for which there is a binary rule, termed scalar multiplication, such that for every $\alpha \in (S, +, \cdot)$ (a scalar) and $x \in \Pi_n$ (a vector) there exists a product $\alpha x \in (\Pi_n, +)$ satisfying the following axioms.

A6. $\alpha(\beta x) = (\alpha \beta)x$ $\forall \alpha, \beta \in S, \quad x \in \Pi_n$
A7. $(\alpha + \beta)x = \alpha x + \beta x$ $\forall \alpha, \beta \in S, \quad x \in \Pi_n$
A8. $\alpha(x + y) = \alpha x + \alpha y$ $\forall \alpha \in S, \quad x, y \in \Pi_n$

A finite set of vector-valued elements of Π_n, $\{x_1, x_2, \ldots, x_n\}$, is *linearly dependent* over $(S, +, \cdot)$ if there exist scalars $\alpha_1, \alpha_2, \ldots, \alpha_n \in S$ *not all zero*, such that

$$\alpha_1 x_1 + \alpha_2 x_2 + \cdots + \alpha_n x_n = 0$$

where 0 denotes the zero vector. If the only relation that exists is the trivial one (that is, $\alpha_i = 0$, $i = 1, 2, \ldots, n$), the set of vectors $\{x_1, x_2, \ldots, x_n\}$ is termed *linearly independent* over $(S, +, \cdot)$.

If $(\Pi_n, +)$ is a vector space, $\Sigma \subset \Pi_n$ is a *spanning set* for Π_n if, for every $x \in \Pi_n$, x may be written as a linear combination of $\alpha_i s_i$, where $\alpha_i \in S$ and $s_i \in \Sigma$. If Σ is finite, then $(\Pi_n, +)$ is a finite-dimensional vector space. Also, $B \subset \Pi_n$ forms a basis for $(\Pi_n, +)$ if B is a minimal spanning set for Π_n.

We state, without proof, the following useful propositions.

1. A basis for a vector space is a linearly independent set.

2. Let B a basis for $(\Pi_n, +)$ over $(S, +, \cdot)$. Then every $x \in \Pi_n$ may be expressed uniquely as a linear combination of elements of B with coefficients in $(S, +, \cdot)$.

3. All bases for a vector space have the same cardinality. Further, the cardinality of a basis for a vector space is equal to the dimension of the space.

3. Inner Product and Normed Spaces

An *inner product*, denoted by $(\cdot, \cdot)$, is a scalar-valued function defined on a vector space X over a field $(S, +, \cdot)$, such that the following axioms are satisfied.

A9. $\forall x, y \in X, \quad (x, y) = (y, x)$

A10. $(\alpha x + \beta y, z) = \alpha(x, z) + \beta(y, z) \qquad \forall x, y, z \in X, \quad \alpha, \beta \in S$

A11. $(x, x) \geqslant 0 \quad \forall x \in X; \qquad (x, x) = 0 \quad \text{iff } x = 0$

Clearly, an inner product is a map that relates the elements of a linear space to the elements of a number field, usually the real or complex numbers. Symbolically, $(\cdot, \cdot): X \to \mathbb{R}$ or $\mathbb{C}$. (A9) illustrates the symmetric nature of an inner product. However, for the $\mathbb{C}$ field the inner product is complex symmetric, and (A9) is rewritten $(x, y) = \overline{(y, x)}$, where the overbar denotes complex conjugation. (A10) describes the inner product as a bilinear functional on X, and (A11) states positive definiteness.

By definition, an inner product space is a vector space with an inner product. Within an inner product space, the *norm* of x, $\forall x \in X$, is the positive square root of (x, x), namely, $\|x\| = (x, x)^{1/2}$.

A norm is a linear functional on X which represents a map from a linear space to the positive real numbers, that is, $\|\cdot\|: X \to \mathbb{R}^+$, such that the following axioms are satisfied.

A12. $\|x\| \geqslant 0 \quad \forall x \in X; \qquad \|x\| = 0 \quad \text{iff } x = 0$

A13. $\|\alpha x\| = \|\alpha\| \|x\| \qquad \forall \alpha \in (S, +, \cdot) \quad x \in X$

A14. $\|x + y\| \leqslant \|x\| + \|y\|$

(A12) states the positive definite nature of a norm, (A13) expresses homogeneity, and (A14) demonstrates subadditivity. By definition, a normed space is a vector space with a norm.

4. Metric Spaces

A *metric* is a scalar-valued distance function, denoted by ρ, which maps $X \times X \rightarrow \mathbb{R}^+$ such that the following axioms are satisfied.

A15. $\rho(x, y) \geqslant 0 \quad \forall x, y \in X; \qquad \rho(x, y) = 0 \quad \text{iff } x = y$

A16. $\rho(x, y) = \rho(y, x)$

A17. $\rho(x, z) \leqslant \rho(x, y) + \rho(y, z)$

(A15) indicates that a metric is strictly positive, (A16) defines a metric as being symmetric, and (A17) is the triangle inequality.

By definition, a metric space is a vector space with a metric. We note that an inner product space has a natural distance function (metric), that is, $\rho(x, y) = \| x - y \| \; \forall x, y \in X$. We also see from the preceding discussion that every inner product space is a normed space, which in turn is a metric space. The converse statement is not true.

References

1. G. Birkhoff and S. MacLane, *A Survey of Modern Algebra*, rev. ed., Macmillan, New York, 1953.

2. A. Clark, *Elements of Abstract Algebra*, Wadsworth, Belmont, CA, 1971.

3. E. T. Copson, *Metric Spaces*, Cambridge University Press, Cambridge, 1968.

4. P. R. Halmos, *Finite-Dimensional Vector Spaces*, 2nd ed., Van Nostrand, Princeton, NJ, 1958.

5. A. P. Hillman and G. L. Alexanderson, *A First Undergraduate Course in Abstract Algebra*, 2nd ed., Wadsworth, Belmont, CA, 1978.

6. I. Kaplansky, *Set Theory and Metric Spaces*, 2nd ed., Chelsea, New York, 1977.

7   Transformations of Macromolecules

Equipped with a basic understanding of DNA structure and its replication, transmission, and expression (Chapter 2), we now proceed to further describe the intricacies underlying the cell's execution of these components of the central dogma of molecular genetics.

1.   Semiconservative Replication of DNA

On the basis of their proposed structure for the DNA molecule, Watson and Crick [1] predicted that each strand of the DNA duplex is employed as a template for the replication of daughter strands which possess the same genetic information (although in complementary notation) as the parent molecule. Thus from a single original DNA molecule, two newly synthesized DNA duplexes result, each of which is comprised of one parental and one daughter strand. Verification of this semiconservative mechanism for DNA replication was provided by Meselson and Stahl [2]. In their experiment, bacterial cells were first allowed to grow for several generations in a medium containing $[^{15}N]H_4Cl$ as its sole source of nitrogen. Subsequently the bacteria were transferred to an $[^{14}N]H_4Cl$ medium, and the fate of the original $[^{15}N]DNA$ was monitored through several replication cycles. Utilizing CsCl equilibrium density gradient centrifugation, the DNA of successive generations was shown to form distinctive sedimentation profiles. Initially a single ^{15}N-labeled heavy DNA band was resolved. After one generation on $[^{14}N]H_4Cl$, a sole hybrid (i.e., one ^{15}N- and one ^{14}N-labeled strand) DNA band was evident. After a second generation, two bands were present. These profiles were the result of the presence of equal amounts of light DNA (^{14}N-labeled only) in addition to the hybrid DNA.

   In bacteria and many viruses, DNA exists as a circular double helix. Replication, in this case, occurs from a single initiation point and proceeds in both directions simultaneously by means of two replication forks [3]. In eukaryotic organisms, however, DNA is localized as nucleosomes associated with chromatin fibers. Bidirectional replication is again the rule, but now literally thousands of origins (i.e., with multiple replication forks) are employed at once so that several eukaryotic chromosomes are reproduced expeditiously [4].

DNA synthesis itself occurs via the mediation of a preexisting DNA duplex (primer strand + template strand) and is catalyzed by DNA polymerase. Kornberg [5–7] demonstrated that, in bacterial extracts, DNA polymerase promotes the covalent addition of successive deoxyribonucleoside 5'-triphosphate molecules by their α-phosphate groups to the free 3'-hydroxyl end of the DNA primer chain in accord with the $(3' \to 5')$ base sequence of the template strand. Thus chain growth for the newly synthesized DNA proceeds in the $5' \to 3'$ direction.

Yet the action of DNA polymerase alone does not constitute the full story for DNA synthesis. Indeed, DNA replication involves a complex set of reaction steps, the complete details of which are still lacking. As it turns out, even DNA polymerase itself is not a single enzyme, but three enzymes designated DNA polymerase I, II, and III. It is DNA polymerase III which is primarily responsible for the chain elongation action described above.

Whereas DNA polymerase may add nucleotides in the $5' \to 3'$ direction only, we must explore some additional synthetic scheme to account for the observation that DNA replication occurs along both (antiparallel) strands of the duplex simultaneously. In fact, only one DNA strand is replicated continuously in the $5' \to 3'$ direction (i.e., the leading strand). The remaining chain (i.e., the lagging strand) is synthesized discontinuously in small sections, termed Okazaki fragments [8]. Okazaki fragments require a complementary RNA primer to which DNA polymerase adds deoxyribonucleotides from the 3' end. Finally, the RNA primer is expunged (by the $5' \to 3'$ exonuclease activity of DNA polymerase I), replaced by complementary DNA, and the fragments are joined together by DNA ligase.

Replication of the tightly wound DNA duplex also requires the action of helicase. This enzyme is capable of unwinding small sections of DNA, just in front of the replication fork, so that the nucleotide base sequence is now exposed and therefore becomes available to the oncoming DNA polymerase system. Since the helix has a natural predilection to immediately rewind, several DNA-binding proteins attach to the separated strands and hold them apart long enough for the DNA polymerase to pass through the region.

The rapid unwinding of the duplex can conceivably create a considerable torque, which might force the entire DNA molecule to rotate wildly in response to such an action. To prevent this, the cell employs a "swivel" mechanism which necessitates that only a short segment of DNA must undergo rotation. The swivel is induced by a class of topoisomerases (e.g., DNA gyrase in prokaryotes) which force a temporary break in one DNA strand that is later repaired.

The DNA replication system is also equipped with a built-in mechanism for proofreading and/or repairing errors introduced into the newly synthesized DNA. DNA polymerase I possesses both $3' \to 5'$ (for proofreading, i.e., removing unpaired nucleotides) and $5' \to 3'$ (for repair of DNA) exonuclease activity.

It is crucial that DNA replication proceed such that the highest degree of fidelity is maintained since any error, however minor, may jeopardize the viability of an organism. In this regard, errors which occur during transcription or translation are potentially better tolerated than replication errors since, in the case of the former, only a single RNA or protein in one cell might be affected, while in the latter circumstance, the entire genetic identity of an organism may be altered [9].

2. DNA Transcription Yields mRNA

In the next stage of genetic information transfer, namely, transcription, the nucleotide sequence of one strand of a DNA duplex serves as a template for the production of a single strand of complementary RNA. Whereas the entire length of chromosome is copied during DNA replication, only those regions of DNA which code for genes are selectively transcribed. The products of transcription are principally the mRNA, tRNA, and rRNA species as well as some specific regulatory sequences. After their synthesis in the nucleus of eukaryotic cells, these RNA species are transported to the cytoplasm where they now become engaged in the business of polypeptide production.

It is in the ribonucleotide sequence of the mRNA molecule that the information necessary to prescribe a specific amino acid sequence in a nascent polypeptide resides. An mRNA may code for one polypeptide (monogenic or monocistronic) or for two or more polypeptides (polygenic or polycistronic). In prokaryotes, polygenic mRNAs also possess intergenic regions which act as nontranslated spacers between mRNA segments specifying different polypeptides.

RNA is synthesized by a DNA-directed RNA polymerase which is capable of catalyzing the addition of ribonucleoside $5'$-triphosphates to the $3'$-hydroxyl end of an RNA strand [10–12]. Thus RNA molecules grow in the $5' \to 3'$ direction and the product RNA possesses opposite polarity with respect to its DNA template. RNA polymerase requires preexisting DNA, and even though only a single chain of the DNA template is transcribed, the enzyme's activity is somewhat higher in the presence of the DNA duplex. In addition, since uracil assumes the role of thymine in the RNA transcript, uracil is added to RNA in response to adenine in the DNA template, that is, complementary base pairing between uracil and adenine exists.

RNA polymerase must bind to a specific initiation point (the promoter site) on the DNA template before transcription may begin. In addition, as successive ribonucleotides are joined to the transcript, the newly synthesized RNA associates with the DNA template to form a hybrid (DNA/RNA) double helix. Such a close association increases the accuracy of transmission of a nucleotide base sequence between the DNA template and the RNA transcript. RNA chain elongation proceeds as the polymerase moves down the DNA template until a specific transcription termination signal is reached. At this point, the RNA transcript falls off the template and transcription is complete.

In eukaryotic organisms, RNA polymerase represents a system of three enzymes, RNA polymerase I, II, and III. RNA polymerase I is involved in rRNA synthesis, RNA polymerase II is responsible for mRNA production, and RNA polymerase III is concerned with tRNA and some rRNA synthesis.

In both prokaryotes and eukaryotes, RNA polymerase produces large precursor rRNAs and tRNAs which are then subjected to posttranscriptional modification (i.e., cleaved by endonucleases and further altered enzymatically) in order to yield the final biologically active tRNA and rRNA species. In the case of eukaryotes, mRNAs are also produced in precursor form as heterogeneous nuclear RNAs (hnRNAs). They too are then subjected to posttranscriptional modification, which includes the addition of an extended poly-A tail and a methylguanosine cap at the 3′ and 5′ ends of the molecule, respectively. More significantly, however, since the hnRNAs are transcribed in a colinear fashion with respect to the DNA template, they contain long stretches of nontranslated intervening sequences, or introns, in addition to the coding sequences, or exons. Therefore in order to obtain the final mRNAs that are observed in the cytoplasm, such introns must be removed. This occurs through the mediation of small nuclear RNAs which selectively splice together neighboring exon regions. The excluded intron segments are then degraded by nucleases while the pure exon-containing product mRNAs are passed to the cytoplasm.

3. mRNAs are Translated into Polypeptide Chains

The sequence of events by which an mRNA is translated into a polypeptide chain is extremely complex. Basically, five separate steps for translation may be distinguished: (1) the activation of amino acids via their highly specified attachment to tRNA, (2) initiation of polypeptide synthesis following the association of mRNA, ribosome, and aminoacyl-tRNA to form an initiation complex, (3) polypeptide chain elongation as a result of the covalent addition of successive amino acids, (4) termination and release of the nascent polypeptide in response to a termination signal (i.e., one of the three termination codons), and (5) the subsequent folding of the polypeptide to establish its biologically active three-dimensional conformation and any ancillary processing of the molecule (e.g., the addition of a prosthetic group).

The tRNAs function as interpretative adaptors between the nucleic acid sequence of the mRNA and the amino acids themselves. As a group, the tRNAs constitute a class of relatively small single-stranded molecules which share several interesting features [13]. A significant number of characteristic, modified nucleic acid bases are found as part of the tRNA primary structure. In addition, all tRNAs share a common -CCA sequence at their 3' end. More importantly for their biological function, if one introduces sufficient hydrogen bonding to allow for intrachain base pairing, the primary structure of tRNAs (i.e., their nucleic acid sequence) now assumes a three-dimensional cloverleaf conformation with four (or five, in some cases) distinct side arms. These include the amino acid arm with a -CCA terminal sequence to which a specific amino acid is esterified, the anticodon arm which contains the anticodon complementary (antiparallel) partner for a specific mRNA codon, the dihydrouridine (DHU) arm, and the TψC (T = ribothymidine, ψ = pseudouridine) arm.

The esterification of an amino acid to its tRNA is mediated by aminoacyl-tRNA synthetases [14]. There are 20 such enzymes present in the cytoplasm of cells, one for each of the 20 amino acids to be incorporated into the growing polypeptide chain. In addition, aminoacyl-tRNA synthetases possess a good deal of specificity with respect to the amino acids and the tRNAs they bind and can in turn proofread their product and subsequently correct any mismatching that has occurred. It is the anticodon of the tRNA which largely determines the specificity with which the aminoacyl-tRNA reacts in response to mRNA.

Polypeptide chains are synthesized from the amino-terminal end (N terminus) to the carboxyl-terminal end (C terminus). In prokaryotes, the first amino acid at the N terminus is always N-formylmethionine (fMet). fMet is carried to the initiation site, via a special tRNA^{fMet}, in the form of N-formylmethionyl-tRNA^{fMet}. (A second tRNA^{Met} inserts methionine as methionyl-tRNA^{Met} along the interior of the polypeptide chain.) In eukaryotic cytoplasmic protein synthesis, the N terminus is occupied by methionine, transported to the initiation site by a specific initiator tRNA^{Met}.*

Ribosomes constitute the physical location for translation. Prokaryotic ribosomes are characterized by a 70 S sedimentation coefficient and are composed of 50 S and 30 S subunits. In the case of eukaryotes, the cytoplasmic

---

*The N-terminal amino acid of polypeptides produced by the mitochondria and chloroplasts of eukaryotic cells is N-formylmethionine. This fact, in concert with other aspects of the protein-synthesizing machinery in chloroplasts and mitochondria, lends support for the idea of an ancestral relationship between these organelles and the bacteria.

ribosomes are 80 S in character and consist of 60 S and 40 S subunits. Ribosomal subunits, when placed together, create a groove through which the mRNA moves during translation [15].

The actual initiation of the translation process proceeds in a series of three steps and requires the smaller of the ribosomal subunits, the mRNA, the initiating N-formylmethionyl-tRNA^{fMet} (N-methionyl-tRNA^{Met} in eukaryotes), a set of three initiation factors (IF-1, IF-2, and IF-3) and GTP (guanosine 5′-triphosphate, a source of cellular energy).* Basically, mRNA binds to the 30 S (40 S in eukaryotes) ribosomal subunit and is brought into position by an initiating signal on the transcript itself (see Figure 2-3). This signal identifies the AUG codon on the 5′ end of the mRNA to which the fMet-tRNA^{fMet} (Met-tRNA^{Met} in eukaryotes) binds. Finally, the larger ribosomal subunit joins the system to form an initiation complex with the concomitant release of the initiation factors and the hydrolysis of GTP $\rightarrow$ GDP $+ P_i$. A fully competent initiation complex includes the mRNA, the 70 S (80 S in eukaryotes) ribosome, and the N-formylmethionyl-tRNA^{fMet} (N-methionyl-tRNA^{Met} in eukaryotes).

All ribosomes possess two distinct sites for binding aminoacyl-tRNAs. The aminoacyl (or A) site accepts incoming aminoacyl-tRNAs with the exception of the initiating one, which binds directly elsewhere. The peptidyl (or P) site holds the nascent peptidyl-tRNA. The actual elongation of the polypeptide chain requires the initiation complex described above, the next aminoacyl-tRNA specified by the codon sequence of the mRNA template, three elongation factors (EF-Tu, EF-Ts, and EF-G), and GTP. The newly specified aminoacyl-tRNA (associated with EF-Tu and GTP) is first bound to the A site on the ribosome. A peptide bond is then formed between the amino acids in the A and P sites by peptidyl transferase. This is accomplished through the transfer of the fMet (Met in eukaryotes) residue from its tRNA on the P site to the amino group of the aminoacyl-tRNA bound to the A site. Next the ribosome shifts down the mRNA (in the 3′ direction) by a distance of one codon (i.e., three nucleotide bases). This translocation step, which requires EF-G and GTP hydrolysis, effectively moves the dipeptidyl-tRNA from the A to the P site, thereby leaving open the A site for the binding of the next aminoacyl-tRNA in sequence.

Polypeptide chain elongation continues in this manner until one of three termination codons is reached in the mRNA template. This effectively signals the completion of protein synthesis. At this point, the polypeptide and its attached tRNA are separated and the ribosome dissociates into its subunits. After synthesis, polypeptide chains are often subjected to posttranslational

---

*For a detailed account concerning the role of the initiation, elongation, and termination processes in translation, see [16].

modification. Processing may include N- or C-terminus alteration, methylation of amino acid R groups, and/or addition of prosthetic groups.

It is of obvious selective advantage for an organism to be able to regulate accurately the rate of protein biosynthesis to meet, but not exceed, its metabolic demands. In bacterial cells, certain enzymes are always manufactured at a constant rate, independent of the metabolic rate. These so-called constitutive enzymes differ from enzymes whose activity may be effectively potentiated or induced in response to the needs of the organism. The activity of other enzymes may be repressed by an overabundance of their end product, for example, in order to conserve cellular resources. In general, regulation of protein synthesis occurs principally at the level of the transcription of a particular DNA to its corresponding product mRNA. However, in eukaryotes at least, some control at the level of translation is also possible.

References

1. J. D. Watson and F. H. C. Crick, *Nature*, **171**, 737; 964 (1953).
2. M. Meselson and F. W. Stahl, *Proc. Natl. Acad. Sci. USA*, **44**, 671 (1958).
3. J. Cairns, *Cold Spring Harbor Symp. Quantum Biol.*, **28**, 43 (1963).
4. H. J. Kriegstein and D. S. Hogness, *Proc. Natl. Acad. Sci. USA*, **71**, 136 (1974).
5. A. Kornberg, *Science*, **131**, 1503 (1960).
6. A. Kornberg, *Cold Spring Harbor Symp. Quantum Biol.*, **43**, 1 (1979).
7. A. Kornberg, *DNA Replication*, Freeman, San Francisco, CA, 1980.
8. R. T. Okazaki, K. Okazaki, K. Sakabe, K. Sugimoto, and A. Sugino, *Proc. Natl. Acad. Sci. USA*, **59**, 598 (1968).
9. A. L. Lehninger, *Principles of Biochemistry*, Worth, New York, 1982.
10. S. B. Weiss, *Proc. Natl. Acad. Sci. USA*, **46**, 1020 (1960).
11. J. Hurwitz, J. J. Furth, M. Anders, P. J. Ortiz, and J. T. August, *Cold Spring Harbor Symp. Quantum Biol.*, **26**, 91 (1961).
12. A. Losick and M. Chamberlin, Eds., *RNA Polymerase*, Cold Spring Harbor Laboratory, Cold Spring Harbor, NY, 1976.
13. A. Rich and S. H. Kim, *Sci. Am.*, **238** (1), 52 (1978).
14. P. R. Schimmel, *Adv. Enzymol.*, **49**, 187 (1979).
15. M. Nomura, A. Tissieres, and P. Lengyl, Eds., *Ribosomes*, Cold Spring Harbor Laboratory, Cold Spring Harbor, NY, 1974.
16. H. Weissbach and S. Pestka, Eds., *Molecular Mechanisms of Protein Biosynthesis*, Academic Press, New York, 1977.

8   Realization of Molecular Genetics in a Linear Space

Prefatory Comments

By exploiting the information contained in the preceding two chapters it now becomes practicable to represent the conceptual nature of molecular genetics in mathematical form. Initially this effects a realization of DNA, RNA, and protein molecules as vectors in separate finite-dimensional vector spaces. As a result, it then becomes feasible to treat transcription and translation as linear operators on certain of these spaces.

Two distinct vector space formulations, different only in field structure, will be developed. The use of a continuous field, specifically the real field, allows for the construction of a familiar Euclidean space. However, this space is considered inadequate to describe the precise linear arrangement of the relevant units (codons or amino acids) in an informational macromolecule. To alleviate this difficulty, a space over a finite field may be created, albeit with the sacrifice of a certain richness of the mathematics. The resolution of this dilemma will come in Chapter 11. Before beginning the vector space constructions, however, we reiterate the fundamental biological sets and the genetic code mapping (see Chapter 5).

The set of four RNA bases is denoted by $B \equiv \{U, A, C, G\}$. Thus the collection of 64 codons, denoted by C, is prescribed by the Cartesian product

$$C \equiv B \times B \times B \qquad (8\text{-}1)$$

At the level of DNA, the base triplets which ultimately define the RNA codons may be similarly specified:

$$C' \equiv B' \times B' \times B' \qquad (8\text{-}2)$$

where $B' \equiv \{T, A, C, G\}$ is the set of DNA bases, and where C' is the collection of DNA codons. Finally, the set of 20 amino acids and the terminator codon (TC) is symbolized by A.

The genetic code mapping is operative during mRNA translation and maps the set C onto the set A. We term this the f mapping, and symbolically

$$f : C \to A \qquad (8\text{-}3)$$

1. Real-Field (ℝ) Construction

Consider the 64-dimensional vector space, denoted by D_1, over the field of real numbers ℝ. Let C' designate a basis for D_1, and further assume that an inner-product is defined on this space. Since C' is a linearly independent set (see Chapter 6, Section 2), it follows that C' may be chosen to be orthogonal.* In addition, let the elements of C' be normalized. Thus,

$$(\alpha_i, \alpha_j) = \delta_{ij} \qquad (8\text{-}4)$$

where $\alpha_i, \alpha_j \in C'$ for $i, j = 1, 2, \ldots, 64$, and where δ_{ij} is the Kronecker delta,

$$\delta_{ij} = \begin{cases} 1 & \text{for } i = j \\ 0 & \text{for } i \neq j \end{cases} \qquad (8\text{-}5)$$

Equation (8-4) determines the structure of D_1 as Euclidean.†
  Since C' spans D_1 (see Chapter 6, Section 2) then for every $c \in D_1$

$$\sum_{i=1}^{64} c_i \alpha_i = c \qquad (8\text{-}6)$$

where $c_1, c_2, \ldots, c_{64} \in ℝ$ are unique. Utilizing Eqs. (8-4) and (8-5), we may state that for every $c, d \in D_1$,

$$(c, d) = \sum_{i=1}^{64} c_i d_i \qquad (8\text{-}7)$$

In view of Eq. (8-7), the 64-tuple $(c_1, c_2, \ldots, c_{64})$ may be considered to be a row vector $\langle c|$ while the 64-tuple $(d_1, d_2, \ldots, d_{64})$ is a column vector $|d\rangle$. $\langle c|$ and $|d\rangle$ are representations of c and d, respectively, in the C' basis.
  The Euclidean norm is well defined on D_1, that is, for every $c \in D_1$,

$$\| c \| = (c, c)^{1/2} = \left[\sum_{i=1}^{64} c_i^2 \right]^{1/2} \qquad (8\text{-}8)$$

---

*This is a well-known result from the theory of linear spaces. The construction process is known as Gram–Schmidt orthogonalization. See, for example [1].
†A Euclidean space is one that is flat and possesses a positive definite metric.

The distance between two vectors $\mathbf{c}, \mathbf{d} \in D_1$ is given by the Euclidean metric

$$\| \mathbf{c} - \mathbf{d} \| = \left[\sum_{i=1}^{64} (c_i - d_i)^2 \right]^{1/2} \tag{8-9}$$

Also, the angle θ between two vectors $\mathbf{c}, \mathbf{d} \in D_1$ may be represented as

$$\theta = \cos^{-1} \left[\frac{(\mathbf{c}, \mathbf{d})^{1/2}}{\| \mathbf{c} \| \cdot \| \mathbf{d} \|} \right] \tag{8-10}$$

Finally, the projection operator is defined as

$$\mathbb{P}_i \mathbf{c} \equiv (\boldsymbol{\alpha}_i, \mathbf{c}) \alpha_i = c_i \boldsymbol{\alpha}_i \tag{8-11}$$

D_1 may be interpreted as a space of DNA vectors where each vector possessing nonnegative, integral coefficients represents a single strand of a given DNA molecule. These nonnegative, integral coefficients c_i represent the number of occurrences of a given DNA codon $\boldsymbol{\alpha}_i$ in a single-stranded DNA (ss-DNA) molecule. Geometrically, c_i is the projection of the DNA vector along the $\boldsymbol{\alpha}_i$ axis in 64-space. Hence the total number of codons n in an ss-DNA may be written as

$$\sum_{i=1}^{64} (\boldsymbol{\alpha}_i, \mathbb{P}_i \mathbf{c}) = \sum_{i=1}^{64} c_i = n \tag{8-12}$$

where $\mathbf{c}$ is the vector realization of the molecule.

By convention we assume that the space D_1 is composed of vectors which represent ss-DNA molecules having $3' \to 5'$ polarity (see Chapter 2, Section 1). An associated space D_1^* constitutes the space of vectors which represent ss-DNA molecules having $5' \to 3'$ polarity. It is obvious, of course, that $D_1 = D_1^*$. In fact, the existence of D_1^* merely illustrates the fact that the opposing strands of a double-stranded DNA (ds-DNA) molecule have opposite polarity with respect to one another. In D_1^* the codons are read from the 5' to the 3' end of the single-stranded molecule in forming the basis set. Thus if $\boldsymbol{\alpha}_i = \text{TAG}$, then $\boldsymbol{\alpha}_i^* = \text{GAT}$. The transformation from D_1 to D_1^* may be viewed as the one-to-one correspondence

$$k: D_1 \to D_1^* \tag{8-13}$$

which simply reverses the polarity of the codon basis vectors.

Now consider the base-antibase mapping g, which is the one-to-one correspondence defined by

$$g(T) = A \qquad g(C) = G$$

$$g(A) = T \qquad g(G) = C$$

(8-14)

Let $c \in D_1$. The operation of g on c is denoted by $g(c) = \bar{c}$. Thus if $\alpha_i = TAC$, then $\bar{\alpha}_i = ATG$.

Utilizing the mappings k and g, we proceed to the realization of a ds-DNA molecule. Let $c \in D_1$ represent the $3' \to 5'$ strand of a given DNA molecule. Let $\mathbf{d}^* \in D_1^*$ represent the $5' \to 3'$ strand of the same molecule. It follows that c and $\mathbf{d}^*$ are related to the same ds-DNA if and only if $c = \bar{\mathbf{d}}^*$ (or, equivalently, $\bar{c} = \mathbf{d}^*$). Thus the total DNA is represented by the pair

$$[c, \mathbf{d}^*]$$

(8-15)

where $c = \bar{\mathbf{d}}^*$.

More than one DNA molecule may be represented by the same vector pair. In fact, a given vector pair represents N distinct DNA molecules, with N being given by

$$N = \frac{n!}{c_1! c_2! \cdots c_{64}!}$$

(8-16)

where n is defined in Eq. (8-12), the c_i are defined in Eq. (8-6), and both n and c_i are nonnegative integers. However, for a given DNA molecule, only one vector pair is defined.

The particular form of Eq. (8-16) results from the manner in which information is encoded in a DNA molecule. This vector formalism adequately describes the *number* of codons of a given type which occurs in a molecule, but the *order* of occurrence is disregarded. Thus a DNA vector pair contains less information than the DNA molecule from which it is formed. This information may be recaptured by a radical restructuring of the field (see Section 2).

A. Transcription

It is now possible to develop a mapping from the space of DNA vector pairs to the space of RNA vectors. This, of course, is the biological process of transcription (see Chapter 7, Section 2).

We postulate the existence of a transcription operator T which operates on a DNA vector pair to produce an RNA vector. This RNA vector represents the biological transcription product of the $3' \to 5'$ DNA vector. Let the DNA vector pair $[\mathbf{c}, \mathbf{d}^*]$ be given. Then

$$T[\mathbf{c}, \mathbf{d}^*] = h(\mathbf{d}^*) = \mathbf{r} \tag{8-17}$$

where $\mathbf{r} \in R_1$, the RNA vector space; and where h is the one-to-one correspondence defined by

$$h: B' \to B \tag{8-18}$$

Specifically,

$$\begin{array}{cc} h(T) = U & h(C) = C \\ \\ h(A) = A & h(G) = G \end{array} \tag{8-19}$$

Note that in Eq. (8-17) $\mathbf{r}$ is defined as having $5' \to 3'$ polarity.

The $5' \to 3'$ vector of the DNA vector pair is transcribed by the operator T^*,

$$T^*[\mathbf{c}, \mathbf{d}^*] = h(k(\mathbf{c})) = \mathbf{r}' \tag{8-20}$$

where $\mathbf{r}' \in R_1$ and k is defined in Eq. (8-13). Again, note that $\mathbf{r}'$ is defined so as to have $5' \to 3'$ polarity. (This is a direct result of the fact that k acts to reverse the polarity of $\mathbf{c}$.) Thus T transcribes the $3' \to 5'$ DNA vector, while T^* transcribes the $5' \to 3'$ vector.

Considered in its totality, the vector space R_1 is isomorphic to D_1. Thus the one-to-one and onto mapping from D_1 to R_1 is

$$h: D_1 \to R_1 \tag{8-21}$$

Hence it is clear that the basis for R_1 is C and that R_1 is structurally identical to D_1. The definition of the operators T and T^*, however, allows attention to be focused directly on a fixed DNA vector pair and its transcription into the correct biologically related RNA vectors. It is for this reason that T and T^* are introduced.

As is the case for D_1, in R_1 the number of RNA molecules corresponding to a given R_1 vector is equal to N, where N is defined in analogy to Eq. (8-16). However, only one R_1 vector corresponds to a given RNA molecule.

B. Translation

As discussed in Chapter 7, Section 3, translation of an mRNA results in the formation of a protein molecule. The rule which assigns a particular amino acid to a given mRNA codon is, of course, that defined by the genetic-code mapping $f: C \rightarrow A$. Protein synthesis is initiated at the AUG codon* of an mRNA molecule and proceeds in the $5' \rightarrow 3'$ direction along the template until one of the terminator codons is reached.

Consider the set of vectors $S_1 \subset R_1$. Here S_1 represents all RNA molecules having an AUG codon at the $5'$ terminus and one and only one terminator codon which is located at the $3'$ terminus. Translation maps S_1 to P_1, the space of protein vectors. P_1 is a 20-dimensional vector space over the field $\mathbb{R}$. The set $A' \equiv A - \{TC\}$ constitutes the basis for P_1. The translation operator is simply the mapping f defined in Eq. (8-3). Thus given $s \in S_1$,

$$f(\mathbf{s}) = \mathbf{p} \qquad (8\text{-}22)$$

where $\mathbf{p} \in P_1$.

Now we designate P_1 as an inner product space with an orthonormal basis. Hence the analogs of Eqs. (8-7)–(8-11) are valid in P_1 space. This permits the length of a protein vector and the distance and angle between two protein vectors to be defined.

The combinatorial formula which enumerates the number of protein molecules related to a fixed vector in S_1 is more complicated than those given previously and will not be discussed explicitly. The complexity results from the constraint placed on the vectors of S_1 and the surjective nature of the mapping f. However, the number N' of protein molecules related to a fixed vector of P_1 is given by

$$N' = \frac{n'!}{p_1! \, p_2! \cdots p_{64}!} \qquad (8\text{-}23)$$

where n' is the total number of amino acids occurring in the protein molecule and the p_i are the expansion coefficients of the protein vector. For the sake of completeness, we state that only one protein vector corresponds to a given protein molecule.

2. Finite-Field (Z_5) Construction

The vector space formulation of Section 1 accounts partially for the total information content of a DNA molecule, that is, individual codons may be

---

*The codon GUG may sometimes act as the initiator codon and is known to code for N-formylmethionine (see [2]). The degeneracy of AUG and GUG as initiator codons may be taken into account by a slight modification of what follows. However, at this time the additional complexity does not appear to be warranted.

distinguished but the linear arrangement of these codons in the DNA molecule is neglected. This statement, of course, also applies to RNA and protein molecules. The total information content contained in the linear arrangement of bases in DNA and RNA may be *explicitly* retained in a vector space structure, however, through the use of a finite field. The field of interest is Z_5 (see Chapter 3, Section 3, and Chapter 6, Sections 1 and 2),

$$Z_5 \equiv \{[0]_5, [1]_5, [2]_5, [3]_5, [4]_5\} \qquad (8\text{-}24)$$

where $[n]_5$ indicates the congruence class of n modulo 5.

We make the order isomorphism,

$$\{[0]_5, [1]_5, [2]_5, [3]_5, [4]_5\} \leftrightarrow \{0, T, A, C, G\} \qquad (8\text{-}25)$$

Thus the field consists of the set $B' \cup \{0\} \equiv \mathscr{H}'$, where 0 is an abstract element which serves as the additive identity. Since in any order isomorphism $[0]_5 \leftrightarrow 0$, there are 24 (not necessarily distinct) order isomorphisms of the form of Eq. (8-25).

Let $\{a_1, a_2, \ldots, a_n\}$ be a set of vectors of cardinality n which is linearly independent over $\mathscr{H}'$. Let D_2 be an n-dimensional vector space over $\mathscr{H}'$ for which $\{a_1, a_2, \ldots, a_n\}$ is a basis. Thus every $c \in D_2$ may be expressed as

$$c = \alpha_1 a_1 + \alpha_2 a_2 + \cdots + \alpha_n a_n \qquad (8\text{-}26)$$

where $\alpha_1, \alpha_2, \ldots, \alpha_n \in \mathscr{H}'$ are unique.

D_2 has the following biological interpretation. Every $c \in D_2$ represents a single-stranded DNA molecule which, by convention, is assumed to have $3' \to 5'$ polarity. The basis vectors simply function as a place-keeping device, while the linear arrangement of DNA bases along the molecule now comprises the ordered set of expansion coefficients of the vector. For example, consider the ss-DNA segment

$$3'\text{-}TACTGGCA\text{-}5'$$

This molecule is represented by the vector $a \in D_2$, where

$$a = Ta_1 + Aa_2 + Ca_3 + Ta_4 + Ga_5 + Ga_6 + Ca_7 + Aa_8$$

In addition, there is an associated space D_2^* which constitutes the vector space representing single-stranded DNA molecules having $5' \to 3'$ polarity. Thus a double-stranded DNA molecule may be delineated as the vector pair

$$[c, d^*] \qquad (8\text{-}27)$$

where $c \in D_2$, $\mathbf{d}* \in D_2^*$, and $\mathbf{c} = \bar{\mathbf{d}}*$. As before, the vector $\bar{\mathbf{d}}*$ is defined by $g(\mathbf{d}*) = \bar{\mathbf{d}}*$, where g is given in Eq. (8-14).

In analogy to Eq. (8-13), let k be the one-to-one correspondence that reverses vector polarity. For example, let $\mathbf{c} \in D_2$ be given as in Eq. (8-26). Then,

$$k(\mathbf{c}) = \mathbf{c}* \tag{8-28}$$

where

$$\mathbf{c}* = \alpha_n \mathbf{a}_1 + \alpha_{n-1} \mathbf{a}_2 + \cdots + \alpha_2 \mathbf{a}_{n-1} + \alpha_1 \mathbf{a}_n \tag{8-29}$$

Obviously, $k(D_2) = D_2^*$.

D_2 is isomorphic to R_2, the vector space representing RNA molecules. This is a field isomorphism prescribed by

$$h: \mathcal{H}' \to \mathcal{H} \tag{8-30}$$

where $\mathcal{H} \equiv \{0, U, A, C, G\}$, and where the one-to-one correspondence h is defined in Eq. (8-18). Implicitly, of course, there exists the order isomorphism

$$\{[0]_5, [1]_5, [2]_5, [3]_5, [4]_5\} \leftrightarrow \{0, U, A, C, G\} \tag{8-31}$$

A. Transcription

The definitions of the transcription operators T and $T*$ are the same as those given previously in Eqs. (8-17) and (8-20), respectively. Thus,

$$T[\mathbf{c}, \mathbf{d}*] = h(\mathbf{d}*) = \mathbf{r} \in R_2 \tag{8-32}$$

and

$$T*[\mathbf{c}, \mathbf{d}*] = h(k(\mathbf{c})) = \mathbf{r}' \in R_2 \tag{8-33}$$

where both $\mathbf{r}$ and $\mathbf{r}'$ have $5' \to 3'$ polarity, and where $\mathbf{r}$ represents the transcription product of the $3' \to 5'$ DNA vector, while $\mathbf{r}'$ represents the transcription product of the $5' \to 3'$ DNA vector.

B. Translation

Since the protein field must contain the set A', it is not possible to form a field set having prime cardinality. To form a finite field, however, the potential field set must be of prime cardinality. Hence a finite field is not defined, and a protein vector space with a structure similar to D_2 and R_2 cannot be constructed. However, the vectors of R_2 may be treated as RNA molecules and utilized to form S_1. The formalism of Section 1.B is then applicable.

3. Discussion

It is now appropriate to assess the two formalisms of molecular genetics just described. Such assessment will be by (1) direct comparison of the two formalisms, (2) indirect comparison through considerations of the utility of the two formalisms in molecular phylogenetic studies, and (3) indirect comparison through considerations of the applicability of the two formalisms to dynamic (evolutionary) questions. Our prime motivation in this section is to direct attention to the biological inadequacies which are inherent in the mathematical formulation presented thus far. This, in turn, will suggest modifications which lead to a faithful mathematical representation of the biological system.

Our concern with molecular phylogenetic studies may be justified as follows: such studies represent an empirical data base for investigations of biological evolution at the molecular level. This gains particular relevance for us, since evolution may be incorporated into the present theory in a natural manner (compare Section 3.C). The end result, however, is the recognition that our Euclidean perspective is truly limiting.

A. *Comparison of the Two Vector Space Formalisms*

In both of the vector space formalisms we have presented, single-stranded DNA molecules and RNA molecules are treated as vectors in finite-dimensional spaces. Double-stranded DNA molecules are treated as vector pairs. Transcription is conceived of as a linear operator mapping DNA space into RNA space. In the first case (Section 1) a protein space is defined and protein synthesis (translation) is treated as a mapping from a subset of RNA space to protein space. An analogous treatment of translation in the second formalism (Section 2) is not possible.

There exist vectors in the real-field construction of Section 1 which do not represent physically realizable molecules, namely, those vectors without nonnegative, integral expansion coefficients. This reflects the fact that this formalism is richer mathematically than the *static* biological system demands. The physically realizable vectors form a countable subset of the space, and the physically unrealizable vectors permit continuous deformations of one physically realizable vector into another. All vectors in the finite-field construction of Section 2, however, represent actual molecules, and in this sense, this latter formalism is a more economical representation of the system.

In Section 1 there is a combinatorial ambiguity in proceeding *from* a vector *to* a molecule, that is, more than one molecule correspond to the same vector. This is a result of the loss of information that occurs when a unique vector is formed from a given molecule. The order of arrangement of the codons in the molecule is not retained in the vector representation. However, Section 2 alleviates this combinatorial ambiguity.

The major drawback of the finite-field construction is that a well-defined Euclidean geometric interpretation does not exist for an arbitrarily dimensioned space. This arises because a positive definite inner product cannot be defined for an arbitrarily dimensioned space.

This is easily seen for a space whose dimension is 5 (or a multiple of 5)— simply form $(\mathbf{v}, \mathbf{v})$ for the vector

$$\mathbf{v} = [1]_5 \mathbf{a}_1 + [1]_5 \mathbf{a}_2 + [1]_5 \mathbf{a}_3 + [1]_5 \mathbf{a}_4 + [1]_5 \mathbf{a}_5$$

where the set $\{\mathbf{a}_1, \mathbf{a}_2, \ldots, \mathbf{a}_5\}$ is assumed to be orthonormal. Such a difficulty does not occur in the real-field representation, and in fact, the existence of a geometric interpretation leads to suggestions for application of this formalism to experimental data (see Section 3.B).

B. Molecular Phylogeny

A considerable number of protein [3] and nucleic acid [4] sequences are available in the literature. Until quite recently, most known nucleic acid sequences were restricted to various types of RNA. However, with the advent of rapid DNA sequencing techniques [5] it is certain that compendia of DNA sequences, from various parts of the genome in diverse organisms, will soon become available. Nevertheless, the great preponderance of molecular sequence data exists for proteins. As a result, a vast literature [6] has arisen concerning the species comparison of the linear amino acid sequence from homologous proteins in an attempt to reconstruct phylogenetic (evolutionary) trees. In this context, homologous proteins possess identical functions and similar physical characteristics in different species (e.g., cytochrome *c*). It is assumed that such proteins are derived from a common ancestor.

The initial work in the area of molecular phylogeny was pioneered by Fitch and Margoliash [7], who compared the protein sequences of cytochrome *c* from 20 different species. These authors defined the *mutation distance* between two cytochromes as the minimal number of nucleotides that must be altered such that the gene for one cytochrome codes for the other. The appropriate mathematics for such a study is that of metric-space theory (see Chapter 6, Section 4). The difficulty with the metric defined by Fitch and Margoliash [7], as has been pointed out previously [8], is that the triangle inequality is not valid. A mathematically correct metric has been proposed [8] which utilizes the concept of the Hausdorff [9] metric in a comparison of all conceivable subsequences of two given sequences. However, it is not possible to prove in general that this function is a metric, and in each case the verification of the triangle inequality necessitated the use of a computer [8].

In the real-field formalism of Section 1 a natural metric exists, namely, the Euclidean metric [see Eq. (8-9)]. However, two difficulties arise in the application of Section 1 to molecular phylogeny, one of which is common to all studies based on protein sequences while the other is intrinsic to the formalism of the vector space.

In all molecular phylogenetic studies predicated on protein sequences, there is an inherent ambiguity in proceeding from a protein to the correct controlling DNA molecule. Such ambiguity is a direct result of the surjective nature of the f mapping. The metric function of Fitch and Margoliash [7] failed mathematically because of the weighting procedure employed in an attempt to correct for the surjective nature of f. As mentioned, however, an appropriate metric may be defined [8], albeit at the cost of extreme complexity. Nevertheless, even in this case it is still necessary to assume that the f mapping is inviolate, and in light of numerous alternative codings, this is not at all certain (see Chapter 5). If, in fact, the genetic concept embodied in the f mapping is, at present, incomplete, then all phylogenetic studies based on protein sequences are suspect. The ideal case, then, would be to utilize DNA sequences directly in molecular phylogeny, but as mentioned above, such data are just now being made available.

The second difficulty encountered in the application of the real-field vector space construction to molecular sequence data is the loss of information implicit in this formalism. Thus distance between two nucleic acid (or protein) sequences, as it is defined here, refers only to differences in codon (or amino acid) composition and neglects the linear arrangement of these information units. To be explicit, if two DNA molecules differ only in codon order, then these two molecular sequences are considered to be at zero distance from one another. In attempting to overcome this artifact, one is driven to employ curved spaces, and this is the motivation for Part 4 of this work.

For completeness we note that the finite-field construction may also be treated as a metric space, and one example of an appropriate metric is that due to Hamming [10]. The Hamming metric simply counts the number of positions in which two sequences differ. Even if we discount our misgivings, mentioned above, concerning phylogenetic studies based on protein sequence data, this formalism is still not applicable to protein sequences (see Section 2.B). Thus phylogenetic analyses utilizing the finite-field vector space must await the availability of a more complete collection of DNA sequences.

C. Dynamic Molecular Genetics

Within the vector space formalism it is a relatively simple matter to define linear operators which act to rotate a vector or to change its length, or to do both. If such operators are functions of time, then they may be interpreted as evolution operators in both a mathematical and a biological sense. At the level of DNA, we may construct operators to account for point mutations (substitutions, insertions, and deletions). Consider the space D_1, for example. DNA mutations may be treated as 64×64 matrices operating on the DNA vectors. Such an extension is equally valid in D_2, although here one deals with $n \times n$ matrices.

In order to explicitly insert dynamics into the present formalism, it is necessary to increment the vector space dimension by 1 to introduce a time coordinate. It is assumed that the values which this coordinate may acquire are all elements of the real line. This then rules out the use of the finite-field construction, since all coordinates would not be over the same field when time is included.* In addition, in light of our previous comments concerning the questionable validity of the f mapping (Section 3.B and Chapter 5), we elect to operate only at the DNA level. What we propose to develop, therefore, is a mathematical theory of molecular evolution, where evolution is presumed to manifest itself through modifications in DNA. The space of interest is D_1 adjoined to a time coordinate, and this is termed the *informational space–time manifold.*

The informational space–time manifold has the cardinality of the continuum. Thus as pointed out previously, there exist vectors in this manifold (at a given time) which do not represent physically realizable DNA molecules. Mathematically, these vectors allow us to posit continuous vector functions and, as such, to employ much richer mathematics. Nevertheless, even though these vectors represent DNA molecules which remain unrealized physically, they are not without biological inference in the case where the codon expansion coefficients are nonintegral but positive. Such vectors may be interpreted as representing an average of related DNA molecules over an entire species, say. Thus rather than being an artifact of the formalism, these DNA vectors offer the possibility of inserting statistics into considerations of the fundamental biological processes of evolution.

The possibility of negative codon expansion coefficients requires additional consideration. In physical spaces, the actual values of the coordinates have no absolute meaning, since it is to changes in coordinates that physical laws apply. In informational space, on the other hand, each point appears to have

---

*One could redefine the finite-field vector space construction at each instant of time, but this would not achieve a mathematical synthesis of the abstract space with time.

absolute meaning as the mathematical representation of either a physically realizable or a physically unrealizable DNA molecule. Thus with statistical considerations in mind, it would seem that, at least for the present, we must reserve biological meaning for only the nonnegative part of informational space. We do so only in the sense that such points (those with nonnegative codon coefficients) have immediate biological interpretation. We do not exclude, however, some future indication of the existence of *biological* laws which might suggest that only changes in codon coordinates have absolute meaning.

D. Further Developments

Now, as a consequence of the preceding discussion, the abstract space of interest is the 65-dimensional informational space–time manifold of DNA vectors with their associated time coordinates. Our understanding of this manifold will be discussed extensively in Chapters 9 and 11. The associated informational space of vectors having opposite polarity need not be considered specifically, since this space is actually identical to the one in question. Likewise, the space of RNA and protein vectors may be generated at any given time by the methods of Section 1 and, hence, need not be explicitly appraised. Thus molecular evolution will now be expressed as motions in the informational space–time manifold.

A persistent problem, which must be rectified, is the biological information loss inherent in informational space–time. This difficulty is alleviated by relaxing our Euclidean constraint and resorting to Riemannian geometries. This, in turn, will foster the concept of a biological evolutionary force. Finally, in Chapter 11 it becomes clear that two DNA sequences which differ only in codon order may possess a nontrivial, nonzero distance between them, when considered as existing at two different evolutionary times. This, however, requires a reevaluation of what is meant by distance.

References

1. P. R. Halmos, *Finite-Dimensional Vector Spaces*, 2nd ed., Van Nostrand, Princeton, NJ, 1958.
2. H. P. Ghosh, D. Söll, and H. G. Khorana, *J. Mol. Biol.*, **25**, 275 (1967).
3. M. O. Dayhoff, *Atlas of Protein Sequence and Structure*, vol. 5, Natl. Biomedical Research Foundation, Silver Springs, MD, 1972; Supplement 1, 1973; Supplement 2, 1976.
4. B. G. Barrell and B. F. C. Clark, *Handbook of Nucleic Acid Sequences*, Joynson-Bruvvers, Oxford, 1974.

5. F. Sanger, G. M. Air, B. G. Barrell, N. L. Brown, A. R. Coulson, J. C. Fiddes, C. A. Hutchinson III, P. M. Slocombe, and M. Smith, *Nature*, **265**, 687 (1977).

6. For a review of this subject, see T. T. Wu, S. M. Fitch, and E. Margoliash, *Ann. Rev. Biochem.*, **43**, 539 (1974); E. Margoliash, *Adv. Chem. Phys.*, **29**, 191 (1975).

7. W. Fitch and E. Margoliash, *Science*, **155**, 279 (1967).

8. W. A. Beyer, M. L. Stein, T. F. Smith, and S. M. Ulam, *Math. Biosci.*, **19**, 9 (1974).

9. F. Hausdorff, *Set Theory*, Chelsea, New York, 1957, p. 166.

10. R. W. Hamming, *Bell Sys. Tech. J.*, **26**, 147 (1950).

PART 4   DYNAMICS

9 Differential Geometry

In this chapter we present certain aspects of local differential geometry (i.e., tensor calculus) which will be exploited in our dynamic formulation of molecular genetics. The introduction of geometries more general than Euclidean geometry reflects our inability, in the formalism of Chapter 8, to define a nonzero distance between vectors for DNA molecules differing only in codon order. However, the definition of a nontrivial, nonzero distance between such DNA molecules, existing at different evolutionary times, becomes practicable if the informational space–time manifold is curved (see Chapter 11). Thus it is requisite that we generalize the results of Chapter 8 to curved spaces.

Our development of tensor calculus will follow the older literature of general relativity [1]. Despite the more recent formulations of differential geometry (i.e., the introduction of coordinate-free methods), we choose to work within the older format for the following reasons.

1. Tensor calculus is an immediate extension of the Euclidean vector space formalism of Chapter 8.
2. There exists an analogy between evolution in informational space–time and gravitation in physical space–time. This analogy derives, in a natural manner, from mathematics: the curvature of the informational space–time manifold (see Section 9) allows for the introduction of a biological evolutionary force.

We restrict our presentation of tensor calculus to only those subjects which have immediate and obvious applications to molecular-genetic concepts. A more complete discussion of this material may be found in the standard references [2].

1. Contravariant and Covariant Vectors [1, 2]

Consider a set of n independent variables $x \equiv \{x^1, x^2, \ldots, x^n\}$ over the real field. Each $x^i \in \mathbb{R}$ may be thought of as the coordinate along the $\mathbf{e}_i$ basis vector in an n-dimensional vector space $\mathbb{R}^n$. $\mathbb{R}^n$ is a particular coordinatization of the space of interest, say M. Consider another set of n variables $x' \equiv \{x'^1, x'^2, \dots, x'^n\}$ over the real field. The x' variables form another coordinatization of M if there exist n independent functions f^i of the x^i variables such that

$$x'^i = f^i(x^1, x^2, \dots, x^n), \qquad i = 1, 2, \dots, n \qquad (9\text{-}1)$$

A necessary and sufficient condition for these functions f^i to be independent is that the Jacobian of the f^i with respect to the x^i not be identically zero [3],

$$\left| \frac{\partial f^i}{\partial x^j} \right| \equiv \begin{vmatrix} \dfrac{\partial f^1}{\partial x^1} & \cdots & \dfrac{\partial f^n}{\partial x^1} \\ \vdots & \ddots & \vdots \\ \dfrac{\partial f^1}{\partial x^n} & \cdots & \dfrac{\partial f^n}{\partial x^n} \end{vmatrix} \neq 0 \qquad (9\text{-}2)$$

When Eq. (9-2) holds, then both the x and x' variables form $\mathbb{R}^n$ coordinatizations of M. If the f^i are C^∞ (read: infinitely differentiable), then M is a differentiable manifold.*

In global terms, the x coordinatization may be thought of as the one-to-one and onto map ϕ,

$$\phi: M \to \mathbb{R}^n[x] \qquad (9\text{-}3)$$

where $\mathbb{R}^n[x]$ denotes the set of n-tuples over the real field in the x coordinatization. The x' coordinatization, on the other hand, is specified by the one-to-one and onto map ϕ',

$$\phi': M \to \mathbb{R}^n[x'] \qquad (9\text{-}4)$$

The connection between ϕ and ϕ' is given by the f^i. Denoting the f^i collectively as f, we have

$$f: \mathbb{R}^n[x] \to \mathbb{R}^n[x'] \qquad (9\text{-}5)$$

---

*The reader should bear in mind that the modern definition of a differentiable manifold is somewhat more complex. The statement given here should not be taken as a definition. For those who would like to pursue this topic further, see [4].

where f is one-to-one and onto. It follows from Eq. (9-2) that the inverse of f exists,

$$f^{-1} \colon \mathbb{R}^n[x'] \to \mathbb{R}^n[x] \tag{9-6}$$

Denoting $f^{-1} \equiv g$, we have, in local terms,

$$x^i = g^i(x'^1, x'^2, \ldots, x'^n), \qquad i = 1, 2, \ldots, n \tag{9-7}$$

Differentiating Eq. (9-7) with respect to x^j yields

$$\frac{\partial x^i}{\partial x^j} = \sum_{k=1}^{n} \frac{\partial g^i}{\partial x'^k} \frac{\partial x'^k}{\partial x^j} = \sum_{k=1}^{n} \frac{\partial x^i}{\partial x'^k} \frac{\partial x'^k}{\partial x^j} \tag{9-8}$$

where the last equality follows from Eq. (9-7). Since the x^i are linearly independent, however, Eq. (9-8) reduces to

$$\sum_{k=1}^{n} \frac{\partial x^i}{\partial x'^k} \frac{\partial x'^k}{\partial x^j} = \delta^i_j \tag{9-9a}$$

where δ^i_j is the Kronecker delta. In a similar manner, we find

$$\sum_{k=1}^{n} \frac{\partial x'^i}{\partial x^k} \frac{\partial x^k}{\partial x'^j} = \delta^i_j \tag{9-10a}$$

It follows from Eq. (9-1) that

$$dx'^i = \sum_{j=1}^{n} \frac{\partial x'^i}{\partial x^j} dx^j, \qquad i = 1, 2, \ldots, n \tag{9-11a}$$

Equation (9-11a) is to be interpreted in the following manner. Consider a point P of M. The differentials dx^i determine any direction at P in the x coordinatization, while the dx'^i determine the same direction in the x' coordinatization.

To simplify the writing of expressions containing summations, we adopt the convention* that whenever an index occurs twice in the same term, a summation over that index is understood. For example, in the summation convention, Eq. (9-11a) becomes

$$dx'^i = \frac{\partial x'^i}{\partial x^j} dx^j \tag{9-11b}$$

---

*This convention is due to Einstein [1] and is referred to by some authors as the Einstein summation convention.

while Eqs. (9-9a) and (9-10a) become

$$\frac{\partial x^i}{\partial x'^k}\frac{\partial x'^k}{\partial x^j} = \delta^i_j \tag{9-9b}$$

and

$$\frac{\partial x'^i}{\partial x^k}\frac{\partial x^k}{\partial x'^j} = \delta^i_j \tag{9-10b}$$

Unless otherwise stated, the index set will always be $\{1, 2, \ldots, n\}$.

A.   Contravariant Vectors

Let A^i be any n functions of the x variables. The A^i are the components of a *contravariant vector* if they obey the same coordinate transformation rule as the dx^i in Eq. (9-11b), that is,

$$A'^j = \frac{\partial x'^j}{\partial x^i} A^i \tag{9-12}$$

Equation (9-12) may be inverted by multiplying both sides by $\partial x^k/\partial x'^j$ and then using Eq. (9-9b),

$$\frac{\partial x^k}{\partial x'^j} A'^j = \frac{\partial x^k}{\partial x'^j}\frac{\partial x'^j}{\partial x^i} A^i = \delta^k_i A^i \tag{9-13}$$

Performing the summation over i on the right-hand side of Eq. (9-13), we find

$$A^k = \frac{\partial x^k}{\partial x'^j} A'^j \tag{9-14}$$

B.   Covariant Vectors

Let A_i be any n functions of the x variables. The A_i are the components of a *covariant vector* if, for any arbitrary contravariant vector B^i,

$$A_i B^i = \text{invariant (scalar)} \tag{9-15}$$

From Eq. (9-15) we may immediately derive the transformation rule for covariant vectors. Since $A_i B^i$ is invariant to coordinate transformations, we have

$$A'_j B'^j = A_i B^i \tag{9-16}$$

Using Eq. (9-14) to invert the B^i, and substituting this result into Eq. (9-16), we find

$$A_j' B'^j = \frac{\partial x^i}{\partial x'^j} A_i B'^j \qquad (9\text{-}17)$$

since the B'^j are arbitrary. However, it follows that

$$A_j' = \frac{\partial x^i}{\partial x'^j} A_i \qquad (9\text{-}18)$$

which is the transformation rule for a covariant vector.

Both contravariant and covariant vectors are order-1 tensors, and their tensor character is defined by the fact that a contravariant order-1 tensor obeys the transformation rule of Eq. (9-12), while a covariant order-1 tensor obeys the transformation rule of Eq. (9-18). A scalar is termed an order-0 tensor and is characterized by its invariance under coordinate transformations.

2. Tensors of Order Greater than 1 [2]

Consider the contravariant vectors A^i, B^i and the covariant vectors A_i, B_i, each vector being represented in the x coordinatization. We set

$$A^{ij} = A^i B^j, \qquad A_{ij} = A_i B_j, \qquad A_j^i = A^i B_j \qquad (9\text{-}19)$$

and denote the same functions in the x' coordinatization by A'^{ij}, A'_{ij}, and $A_j'^i$. From Eqs. (9-12) and (9-18) we have

$$A'^{ij} = \frac{\partial x'^i}{\partial x^k} \frac{\partial x'^j}{\partial x^l} A^{kl} \qquad (9\text{-}20a)$$

$$A'_{ij} = \frac{\partial x^k}{\partial x'^i} \frac{\partial x^l}{\partial x'^j} A_{kl} \qquad (9\text{-}20b)$$

$$A_j'^i = \frac{\partial x'^i}{\partial x^k} \frac{\partial x^l}{\partial x'^j} A_l^k \qquad (9\text{-}20c)$$

Any two sets of functions, possessing different coordinatizations, which satisfy one of the transformation rules [Eqs. (9-20a–c)] is termed a tensor of order 2 and is one of the following types:

The A^{ij} form a contravariant tensor of order 2.

The A_{ij} form a covariant tensor of order 2.

The A^i_j form a mixed tensor of order 2.

On the other hand, not every tensor of order 2 can necessarily be obtained from vectors as in Eq. (9-19). In fact, any n^2 functions of the x variables may be taken as an order-2 tensor (of any type) in the x coordinatization; the appropriate relation [Eqs. (9-20a–c)] may then be utilized to define the components of the same tensor in the x' coordinatization.

The definition of tensors of any order and type results from a generalization of Eqs. (9-20a–c). Thus we have

$$A'^{j_1 j_2 \cdots j_m} = \frac{\partial x'^{j_1}}{\partial x^{k_1}} \frac{\partial x'^{j_2}}{\partial x^{k_2}} \cdots \frac{\partial x'^{j_m}}{\partial x^{k_m}} A^{k_1 k_2 \cdots k_m} \tag{9-21a}$$

$$A'_{j_1 j_2 \cdots j_m} = \frac{\partial x^{k_1}}{\partial x'^{j_1}} \frac{\partial x^{k_2}}{\partial x'^{j_2}} \cdots \frac{\partial x^{k_m}}{\partial x'^{j_m}} A_{k_1 k_2 \cdots k_m} \tag{9-21b}$$

$$A'^{j_1 j_2 \cdots j_m}_{p_1 p_2 \cdots p_r} = \frac{\partial x'^{j_1}}{\partial x^{k_1}} \frac{\partial x'^{j_2}}{\partial x^{k_2}} \cdots \frac{\partial x'^{j_m}}{\partial x^{k_m}} \frac{\partial x^{q_1}}{\partial x'^{p_1}}$$

$$\cdot \frac{\partial x^{q_2}}{\partial x'^{p_2}} \cdots \frac{\partial x^{q_r}}{\partial x'^{p_r}} A'^{k_1 k_2 \cdots k_m}_{q_1 q_2 \cdots q_r} \tag{9-21c}$$

Equation (9-21a) defines a contravariant tensor of order m, Eq. (9-21b) defines a covariant tensor of order m, and Eq. (9-21c) defines a tensor of order $(m + r)$ which is contravariant of order m and covariant of order r. If we denote a tensor by the ordered pair (a, b) (where a denotes the number of contravariant indices and b the number of covariant indices), the total order of the tensor is then $(a + b)$.

A.   Symmetric Tensors

Consider the type $(2, 0)$ tensor A^{ij}. This tensor is *symmetric* if $A^{ij} = A^{ji}$ for all components. A tensor of any order and type is symmetric with respect to two fixed indices (either contravariant or covariant) if the interchange of these indices does not change the tensor. A tensor which is symmetric with respect to all contravariant and all covariant pairs of indices is simply termed symmetric.

In general, an order-2 tensor has n^2 independent components. If the tensor is symmetric, however, there are only $n(n + 1)/2$ independent components.

A tensor which is symmetric with respect to two or more indices in one coordinatization is symmetric in all coordinatizations. This may be illustrated as follows. Let $A^{k_1 k_2 \cdots k_m}$ be a type $(m, 0)$ tensor which is symmetric with respect to k_1, k_2. It follows from Eq. (9-21a), then, that

$$A'^{j_1 j_2 \cdots j_m} = \frac{\partial x'^{j_1}}{\partial x^{k_1}} \frac{\partial x'^{j_2}}{\partial x^{k_2}} \cdots \frac{\partial x'^{j_m}}{\partial x^{k_m}} A^{k_1 k_2 \cdots k_m}$$

$$= \frac{\partial x'^{j_1}}{\partial x^{k_1}} \frac{\partial x'^{j_2}}{\partial x^{k_2}} \cdots \frac{\partial x'^{j_m}}{\partial x^{k_m}} A^{k_2 k_1 \cdots k_m}$$

$$= A'^{k_2 k_1 \cdots k_m} \tag{9-22}$$

B. Antisymmetric Tensors

Consider the type $(2, 0)$ tensor A^{ij}. This tensor is *antisymmetric* if $A^{ij} = - A^{ji}$ for all components. A tensor of any order and type is antisymmetric with respect to fixed indices (either contravariant or covariant) if the interchange of these indices results only in the sign change of the tensor components. A tensor that is antisymmetric with respect to all contravariant and all covariant pairs of indices is simply termed antisymmetric.

If A^{ij} is antisymmetric, then $A^{ii} = - A^{ii} = 0$. Thus an order-2 antisymmetric tensor has only $n(n-1)/2$ independent components.

As was the case for a symmetric tensor, if a tensor is antisymmetric with respect to two or more pairs of indices in one coordinatization, then it is antisymmetric in all coordinatizations.

3. Tensor Addition, Subtraction, Outer Multiplication, Contraction, and Inner Multiplication [1, 2]

A. Addition and Subtraction

The sum or difference of two tensors, of the same order and type, provides a tensor of the same order and type. This follows immediately from Eqs. (9-21a–c).

Consider the type $(2, 0)$ tensor A^{ij}. This tensor may be expressed as

$$A^{ij} = \tfrac{1}{2}(A^{ij} + A^{ji}) + \tfrac{1}{2}(A^{ij} - A^{ji}) \tag{9-23}$$

where the first term on the right-hand side is a symmetric tensor and the second term is antisymmetric. In general, any contravariant (or covariant) tensor of order 2 may be written as the sum of a symmetric and an antisymmetric tensor of order 2.

B. Outer Multiplication

The process of combining two vectors to give a tensor of order 2 (see Section 2) may be generalized to join tensors of any order. The result is a tensor whose order is the sum of the orders of the original tensors. For example, let A^{ij} and B_{rst} be tensors in the x coordinatization. From Eqs. (9-21a, b) we have

$$A'^{kl} B'_{uvw} = \frac{\partial x'^{k}}{\partial x^{i}} \frac{\partial x'^{l}}{\partial x^{j}} \frac{\partial x^{r}}{\partial x'^{u}} \frac{\partial x^{s}}{\partial x'^{v}} \frac{\partial x^{t}}{\partial x'^{w}} A^{ij} B_{rst} \qquad (9\text{-}24)$$

Thus $A^{ij} B_{rst}$ is a tensor of type $(2, 3)$.

In general, then, the multiplication of the components of any number of tensors results in a tensor (the outer product) which is contravariant and covariant to the order obtained by adding the contravariant and covariant indices, respectively, of the original tensors.

C. Contraction

From a mixed tensor of order m, a tensor of order $m - 2$ is obtained by equating one contravariant and one covariant index and then summing over this index. For example, consider the type $(2, 3)$ tensor A^{ij}_{rst}. A^{ij}_{rsj} is a sum of n terms and is a type $(1, 2)$ tensor. The tensor character is demonstrated by

$$\begin{aligned} A'^{kl}_{uvl} &= \frac{\partial x'^{k}}{\partial x^{i}} \frac{\partial x'^{l}}{\partial x^{j}} \frac{\partial x^{r}}{\partial x'^{u}} \frac{\partial x^{s}}{\partial x'^{v}} \frac{\partial x^{t}}{\partial x'^{l}} A^{ij}_{rst} \\[2mm] &= \frac{\partial x'^{k}}{\partial x^{i}} \frac{\partial x^{r}}{\partial x'^{u}} \frac{\partial x^{s}}{\partial x'^{v}} \delta^{t}_{j} A^{ij}_{rst} \\[2mm] &= \frac{\partial x'^{k}}{\partial x^{i}} \frac{\partial x^{r}}{\partial x'^{u}} \frac{\partial x^{s}}{\partial x'^{v}} A^{ij}_{rsj} \end{aligned} \qquad (9\text{-}25)$$

It follows, then, from Eq. (9-21c) that A^{ij}_{rsj} is a tensor of type $(1, 2)$.

The process described above is termed *contraction*. When a mixed tensor is contracted repeatedly, the order of the resultant tensor is always decreased (for each contraction) by 2.

D. Inner Multiplication

The combination of outer multiplication and contraction may also be exploited to compose tensors. For example, from the tensors A^{ij} and B_{rst} we may construct a type $(1, 2)$ tensor such as $A^{ij} B_{jst}$, or a covariant vector such as

$A^{ij}B_{ijt}$. The process of forming a tensor by this method is *inner multiplication*, and the resultant tensor is termed an *inner product*.* In fact, we previously utilized inner multiplication in Eq. (9-15).

We next prove a proposition [1, p. 126] that is a special case of the more general *quotient law* of tensors [2, p. 14]. Let A_{ij} and B^{kl} be tensors. As explained above, the inner product $A_{ij}B^{ij}$ is a scalar. However, assuming a contrary stance, if $A_{ij}B^{ij}$ is a scalar for any choice of the tensor B^{ij}, then A_{ij} is a tensor. The proof of this statement results from the hypothesis

$$A'_{kl}B'^{kl} = A_{ij}B^{ij} \tag{9-26}$$

But since B^{ij} is a tensor, the inversion of Eq. (9-20a) gives

$$B^{ij} = \frac{\partial x^i}{\partial x'^k} \frac{\partial x^j}{\partial x'^l} B'^{kl} \tag{9-27}$$

Substituting Eq. (9-27) into Eq. (9-26) yields

$$\left(A'_{kl} - \frac{\partial x^i}{\partial x'^k} \frac{\partial x^j}{\partial x'^l} A_{ij} \right) B'^{kl} = 0 \tag{9-28}$$

Since the tensor B'^{kl} is arbitrary, the quantity in parentheses must vanish. The result then follows from Eq. (9-20b). An analogous proof may be constructed for tensors of any type and order. Indeed, this theorem was exploited in the derivation of the transformation rule for a covariant vector [Eqs. (9-15)–(9-18)].

4. Order-2 Conjugate Symmetric Tensors [2]

Consider the type $(0, 2)$ symmetric tensor g_{ij}.[†] We require that the determinant formed from the g_{ij} is nonzero, and we denote this determinant by g. If g^{ij} represents the cofactor of the component g_{ij} divided by g, then, from the elementary theory of determinants,

$$g^{kj}g_{ij} = \delta^k_i \tag{9-29}$$

---

*This is a generalization of the inner product defined in Chapter 6, Section 3. A11 need not hold in this case, however.

[†]The condition that g_{ij} be symmetric is required for the development of the process of raising and lowering indices [see Eqs. (9-35a, b)].

Let A^i be an arbitrary contravariant vector. Then $g_{ij}A^i$ is an arbitrary covariant vector, say B_j. From Eq. (9-29),

$$g^{kj}B_j = g^{kj}g_{ij}A^i = \delta_i^k A^i = A^k \tag{9-30}$$

Utilizing a generalization of the reasoning presented in Section 3.D, g^{kj} may be shown to possess tensor character. Since A^k is a contravariant vector, then, from Eqs. (9-30) and (9-12),

$$g'^{kj}B_j' = \frac{\partial x'^k}{\partial x^l} g^{li} B_i \tag{9-31}$$

The inversion of Eq. (9-18) gives

$$B_i = \frac{\partial x'^j}{\partial x^i} B_j' \tag{9-32}$$

Substituting Eq. (9-32) into Eq. (9-31) yields

$$\left(g'^{kj} - \frac{\partial x'^k}{\partial x^l} \frac{\partial x'^j}{\partial x^i} g^{li} \right) B_j' = 0 \tag{9-33}$$

Since B_j' is arbitrary, however, the quantity in parentheses vanishes, and the tensor character of g^{ij} follows from Eq. (9-20a). Consequently, we may state:

If g_{ij} is a symmetric tensor of type $(0, 2)$ whose (nonzero) determinant is denoted by g, then the cofactors of the g_{ij} divided by g constitute a symmetric tensor g^{ij} of type $(2, 0)$.

Clearly, if the tensor g^{ij} was initially assumed, the tensor g_{ij} may be derived in a manner analogous to that above. Hence g^{ij} and g_{ij} are the *conjugate* of each other.

    Contraction of Eq. (9-29) results in the invariant δ_i^i, which is simply the trace of the n-dimensional unit matrix. Therefore,

$$g^{ij}g_{ij} = n \tag{9-34}$$

Incidentally, Eq. (9-29) proves the tensor character of the Kronecker delta: δ_k^i is a tensor of type $(1, 1)$.

    With the methods presented in Section 3 a symmetric tensor g_{ij}, along with its conjugate g^{ij}, can be exploited to generate tensors of the same order, but different type, from a given tensor. Let A_{ijk} be a type $(0, 3)$ tensor. Then, for example,

$$g^{il}A_{ijk} = A^l{}_{jk}, \qquad g^{jl}A_{ijk} = A_i{}^l{}_k \qquad (9\text{-}35\text{a})$$

and so on. From a type $(3,0)$ tensor A^{ijk},

$$g_{il}A^{ijk} = A_l{}^{jk}, \qquad g_{jl}A^{ijk} = A^i{}_l{}^k \qquad (9\text{-}35\text{b})$$

and so on. The process illustrated by Eqs. (9-35a, b) constitutes the raising and lowering of indices, respectively. Tensors generated in this manner are said to be *associate* to the original tensor by g_{ij}. When dealing with associate tensors, our original notation for writing indices is modified such that the original index location is marked by a gap [see Eqs. (9-35a, b)].

5. The Fundamental Tensor [2]

The differential element of length ds (the metric) in a three-dimensional Euclidean space over the real field (in Cartesian coordinates) is given by

$$ds^2 = (dx^1)^2 + (dx^2)^2 + (dx^3)^2 \qquad (9\text{-}36)$$

This idea, generalized by Riemann [5], defines the differential element of length in an n-dimensional space by*

$$ds^2 = g_{ij}\,dx^i\,dx^j \qquad (9\text{-}37)$$

where $g_{ij} = g_{ij}(x^1, x^2, \ldots, x^n)$ and $g = |g_{ij}| \neq 0$. Historically only spaces in which Eq. (9-37) is positive definite were studied, but with the advent of the general theory of relativity, this restriction was relaxed.† In modern terminology, a space having a positive definite metric‡ of the form of Eq. (9-37) is termed *Riemannian*. A geometry based on such a metric is a *Riemannian geometry*.

Following Section 3.D we state that since ds must be a differential invariant, and in as much as the contravariant differential vectors dx^i, dx^j are necessarily arbitrary, g_{ij} is a tensor of type $(0, 2)$ which we assume, without loss of

---

*We assume that all spaces are over the real field.
†The metric of general relativity is indefinite.
‡We take the positive square root of ds^2 in forming ds.
$^\|$In what follows we shall assume that the metric is always positive definite. For the removal of this restriction, see [2, p. 35].

generality, to be symmetric. Thus ds^2 is the inner product of g_{ij}, with the outer product of the two contravariant differential vectors dx^i, dx^j. The differential element of length may be (loosely) thought of as the *magnitude* of the contravariant differential vector dx^i. Thus by the methods of Section 4, g_{ij} is such that

$$g_{ij} dx^i = dx_j \tag{9-38}$$

Hence Eq. (9-37) may also be written

$$ds^2 = dx_j dx^j \tag{9-39}$$

In general, let a contravariant vector A^i be given. The squared magnitude of this vector then is

$$A^2 = g_{ij} A^i A^j = A_j A^j \tag{9-40}$$

Reasoning from Eq. (9-36) then, the Euclidean character of an n-dimensional space is totally specified by choosing the fundamental tensor to be the unit matrix in n dimensions. Thus it follows from Eq. (9-38) that covariant and contravariant vectors are identical in a Euclidean space.

6. Geodesics [2]

Let P_1 and P_2 be any two points in the n-dimensional real continuum. Let C be a curve defined by $x^i = f^i(\tau)$, where τ is any real parameter such that P_1 and P_2 are points of C with parametric values τ_1 and τ_2, respectively. Consider the n equations

$$\bar{x}^i = x^i + \varepsilon \omega^i \tag{9-41}$$

where ε is an infinitesimal and the ω^i are defined as functions of the x^i. The ω^i are chosen such that

$$\omega^i = 0 \quad \text{when} \quad \tau = \tau_1, \tau_2 \tag{9-42}$$

Equations (9-41) and (9-42) define a curve $\bar{C}$ which contains P_1 and P_2 and which lies infinitesimally close to C.

Let $\dot{x}^i = dx^i/d\tau$, and let φ be an analytic function of the $2n$ variables $x^i, \dot{x}^i$. Consider the integral

$$I = \int_{\tau_1}^{\tau_2} \varphi(x^1, x^2, \ldots, x^n, \dot{x}^1, \dot{x}^2, \ldots, \dot{x}^n) \, d\tau \tag{9-43}$$

Denoting the analogous integral for the curve by $\bar{C}$ and $\bar{I}$, we have, by a Taylor series expansion of φ,

$$\bar{I} - I = \varepsilon \int_{\tau_1}^{\tau_2} \left[\frac{\partial \varphi}{\partial x^i} \omega^i + \frac{\partial \varphi}{\partial \dot{x}^i} \dot{\omega}^i \right] d\tau + \cdots \tag{9-44}$$

where $\dot{\omega}^i = (\partial \omega^i / \partial x^j) \dot{x}^j$, and where only the term that is first order in ε is explicitly written. Thus for a first-order variation in I, holding the end points of the curve fixed [Eq. (9-42)],

$$\delta I = \varepsilon \int_{\tau_1}^{\tau_2} \left[\frac{\partial \varphi}{\partial x^i} \omega^i + \frac{\partial \varphi}{\partial \dot{x}^i} \dot{\omega}^i \right] d\tau \tag{9-45}$$

Integration by parts of the second term in Eq. (9-45) yields

$$\int_{\tau_1}^{\tau_2} \frac{\partial \varphi}{\partial \dot{x}^i} \dot{\omega}^i \, d\tau = \omega^i \frac{\partial \varphi}{\partial \dot{x}^i} \Big|_{\tau_1}^{\tau_2} - \int_{\tau_1}^{\tau_2} \frac{d}{d\tau} \left(\frac{\partial \varphi}{\partial \dot{x}^i} \right) \omega^i \, d\tau \tag{9-46}$$

The first term in Eq. (9-46) vanishes because of Eq. (9-42). Thus Eq. (9-45) becomes

$$\delta I = \varepsilon \int_{\tau_1}^{\tau_2} \left[\frac{\partial \varphi}{\partial x^i} - \frac{d}{d\tau} \left(\frac{\partial \varphi}{\partial \dot{x}^i} \right) \right] \omega^i \, d\tau \tag{9-47}$$

If δI vanishes for any set of functions ω^i which satisfy Eq. (9-42), but are otherwise arbitrary, then I is *stationary*, and C is *extremal* (not necessarily minimal). From Eq. (9-47), then, a necessary and sufficient condition that I be stationary is that

$$\frac{\partial \varphi}{\partial x^i} - \frac{d}{d\tau} \left(\frac{\partial \varphi}{\partial \dot{x}^i} \right) = 0$$

or, by a change of sign,

$$\frac{d}{d\tau} \left(\frac{\partial \varphi}{\partial \dot{x}^i} \right) - \frac{\partial \varphi}{\partial x^i} = 0 \tag{9-48}$$

Equations (9-48) constitute *Euler's equations*.

Now consider the application of this general method to the special case when

$$\varphi = \left[g_{ij} \frac{dx^i}{d\tau} \frac{dx^j}{d\tau} \right]^{1/2} \equiv [g_{ij}\dot{x}^i\dot{x}^j]^{1/2} \tag{9-49}$$

Thus the integral in question is [see Eq. (9-37)]

$$s = \int_{\tau_1}^{\tau_2} [g_{ij}\dot{x}^i\dot{x}^j]^{1/2} \, d\tau \tag{9-50}$$

Of interest is the derivation of the differential equations for a curve with stationary *arc length s*. Such a curve is termed a *geodesic* and constitutes the generalization of the Euclidean straight line.

Using Eq. (9-49),

$$\frac{\partial\varphi}{\partial\dot{x}^i} = \frac{g_{ij}\dot{x}^j}{[g_{ij}\dot{x}^i\dot{x}^j]^{1/2}} = \frac{g_{ij}\dot{x}^j}{ds/d\tau} \tag{9-51}$$

and

$$\frac{\partial\varphi}{\partial x^i} = \frac{1}{2}\frac{(\partial g_{jk}/\partial x^i)\dot{x}^j\dot{x}^k}{[g_{jk}\dot{x}^j\dot{x}^k]^{1/2}} = \frac{1}{2}\frac{(\partial g_{jk}/\partial x^i)\dot{x}^j\dot{x}^k}{ds/d\tau} \tag{9-52}$$

To simplify the writing of the following equations, we adopt the notational convention

$$\frac{\partial g_{jk}}{\partial x^i} = g_{jk,i} \tag{9-53}$$

Now, substituting Eqs. (9-51) and (9-52) into Eq. (9-48) yields

$$g_{ij}\ddot{x}^j + g_{ij,k}\dot{x}^j\dot{x}^k - \frac{1}{2}g_{jk,i}\dot{x}^j\dot{x}^k - g_{ij}\dot{x}^j\left[\frac{d^2s/d\tau^2}{ds/d\tau}\right] = 0 \tag{9-54}$$

which may be rewritten as

$$g_{ij}\frac{d^2x^j}{d\tau^2} + \Gamma_{ijk}\frac{dx^j}{d\tau}\frac{dx^k}{d\tau} - g_{ij}\frac{dx^j}{d\tau}\left[\frac{d^2s/d\tau^2}{ds/d\tau}\right] = 0 \tag{9-55}$$

In Eq. (9-55) Γ_{ijk} is the *Christoffel symbol of the first kind*, which is defined as

$$\Gamma_{ijk} \equiv \tfrac{1}{2}(g_{ij,k} + g_{ik,j} - g_{jk,i}) \tag{9-56}$$

Multiplication of Eq. (9-55) by g^{il} and summation over i gives

$$\frac{d^2x^l}{d\tau^2} + \Gamma^l_{jk}\frac{dx^j}{d\tau}\frac{dx^k}{d\tau} - \frac{dx^l}{d\tau}\left[\frac{d^2s/d\tau^2}{ds/d\tau}\right] = 0 \qquad (9\text{-}57)$$

where Γ^l_{jk} is the *Christoffel symbol of the second kind*, defined as

$$\Gamma^l_{jk} \equiv g^{il}\Gamma_{ijk} \qquad (9\text{-}58)$$

Finally, if the arc length s of the curve is used instead of the completely general parameter τ, Eq. (9-57) becomes*

$$\frac{d^2x^l}{ds^2} + \Gamma^l_{jk}\frac{dx^j}{ds}\frac{dx^k}{ds} = 0 \qquad (9\text{-}59)$$

The extremals of Eq. (9-50), where the parameter τ is taken to be the arc length s, are the integral curves of the n ordinary differential equations [Eq. (9-59)]. These curves satisfy the condition that, anywhere along the curve,

$$g_{ij}\frac{dx^i}{ds}\frac{dx^j}{ds} = 1 \qquad (9\text{-}60)$$

which follows from Eq. (9-50). Equation (9-60) expresses the fact that the length of the vector tangent to the curve is the same at every point on the curve.

Certain properties of Christoffel symbols are immediately evident. From Eqs. (9-56) and (9-58) it is clear that Γ_{ijk} and Γ^i_{jk} are *not* tensors (see Section 7).

In addition, Eq. (9-56) demonstrates that Γ_{ijk} is symmetric in the last two indices. Using this fact and Eq. (9-58), Γ^i_{jk} is shown to be symmetric in the two lower indices. Also, from Eq. (9-56) it follows that

$$\Gamma_{ijk} + \Gamma_{jik} = g_{ij,k} \qquad (9\text{-}61)$$

7. Covariant Differentiation [1, p. 133]

Let ϕ be a scalar and form the gradient of ϕ with respect to the x^i

$$\frac{\partial\phi}{\partial x^i} \equiv \phi_{,i} \qquad (9\text{-}62)$$

---

*In a space in which curves of zero length can exist, these curves are not described by Eq. (9-56). Since we have chosen a positive definite metric, however, such curves do not exist.

Under a change of coordinatization, $\phi_{,i}$ transforms as

$$\frac{\partial \phi}{dx'^j} = \frac{\partial \phi}{\partial x^i} \frac{\partial x^i}{\partial x'^j} \tag{9-63}$$

which may be rewritten as

$$\phi'_{,j} = \frac{\partial x^i}{\partial x'^j} \phi_{,i} \tag{9-64}$$

Comparison of Eq. (9-64) with Eq. (9-18) indicates that $\phi_{,i}$ is a covariant vector.

Consider the covariant vector A_k. To ascertain whether or not its derivative $A_{k,l}$ is a tensor, we proceed as follows. From Eq. (9-18), A_k transforms as

$$A'_i = \frac{\partial x^k}{\partial x'^i} A_k \tag{9-65}$$

Hence,

$$
\begin{aligned}
A'_{i,j} &= \left(\frac{\partial x^k}{\partial x'^i} A_k \right)_{,j} \\
&= \frac{\partial x^k}{\partial x'^i} \frac{\partial A_k}{\partial x^l} \frac{\partial x^l}{\partial x'^j} + A_k \frac{\partial^2 x^k}{\partial x'^i \partial x'^j} \\
&= \frac{\partial x^k}{\partial x'^i} \frac{\partial x^l}{\partial x'^j} A_{k,l} + A_k \frac{\partial^2 x^k}{\partial x'^i \partial x'^j} \tag{9-66}
\end{aligned}
$$

Thus $A_{k,l}$ is not a tensor due to the second term on the right-hand side of Eq. (9-66). In general, differentiation of a tensor yields another tensor only when the original tensor is a scalar. It is advantageous to develop an alternative approach such that when a tensor is differentiated, a new tensor is obtained. This process is known as *covariant differentiation*.

Let ϕ be a scalar and $\phi_{,i}$ its derivative. Then,

$$\psi \equiv \frac{d\phi}{ds} = \phi_{,i} \frac{dx^i}{ds} \tag{9-67}$$

where s is the arc length of a given curve. From Eq. (9-67) it follows that ψ is an invariant. Further, ψ is an invariant for all curves starting from a given point, that is, for any choice of the vector dx^i. Now since s is an invariant,

$$\chi = \frac{d\psi}{ds}$$ (9-68)

is also an invariant. Substituting Eq. (9-67) into Eq. (9-68) gives

$$\chi = \frac{d^2\phi}{dx^i\,dx^j}\frac{dx^i}{ds}\frac{dx^j}{ds} + \frac{d\phi}{dx^i}\frac{d^2x^i}{ds^2}$$

$$= \phi_{,i,j}\frac{dx^i}{ds}\frac{dx^j}{ds} + \phi_{,i}\frac{d^2x^i}{ds^2}$$ (9-69)

If the curve with respect to which we have differentiated is taken to be a geodesic, then substitution of (d^2x^i/ds^2) from Eq. (9-59) into Eq. (9-69) yields

$$\chi = \phi_{,i,j}\frac{dx^i}{ds}\frac{dx^j}{ds} - \Gamma^i_{jk}\phi_{,i}\frac{dx^j}{ds}\frac{dx^k}{ds}$$ (9-70)

With an appropriate change of notation, Eq. (9-70) becomes

$$\chi = (\phi_{,j,k} - \Gamma^i_{jk}\phi_{,i})\frac{dx^j}{ds}\frac{dx^k}{ds}$$ (9-71)

Now since a geodesic curve may be traced in any direction from a point, it follows that the ratio of the components of (dx^j/ds) is arbitrary. Hence by the methods of Section 3.D, the quantity in parentheses in Eq. (9-71) is a type of $(0, 2)$ tensor.

Therefore the following result is obtained. Starting from the covariant tensor of order 1,

$$A_j = \frac{\partial\phi}{\partial x^j}$$ (9-72)

where ϕ is a scalar, differentiation provides a covariant tensor of order 2,

$$A_{j;k} = A_{j,k} - \Gamma^i_{jk}A_i$$ (9-73)

which is termed the *covariant derivative* of A_j. (The semicolon before a lower index denotes a covariant derivative just as a comma denotes a regular derivative.)

It is necessary to prove [1, p. 134] that Eq. (9-73) defines a tensor whether or not A_j results from a gradient of a scalar. Let ψ and ϕ be scalars. Then $\psi(\partial\phi/\partial x^j)$ is a covariant vector. Defining

$$S_j \equiv \psi^{(1)}\frac{\partial\phi^{(1)}}{\partial x^j} + \psi^{(2)}\frac{\partial\phi^{(2)}}{\partial x^j} + \cdots + \psi^{(n)}\frac{\partial\phi^{(n)}}{\partial x^j} \tag{9-74}$$

S_j is also a covariant vector if $\psi^{(1)}, \phi^{(1)}, \dots, \psi^{(n)}, \phi^{(n)}$ are scalars. However, *any* covariant vector A_j may be represented in this manner. If the components of A_j are any given functions of the x^i, then

$$\psi^{(1)} = A_1, \qquad \phi^{(1)} = x^1$$
$$\psi^{(2)} = A_2, \qquad \phi^{(2)} = x^2$$
$$\vdots$$
$$\psi^{(n)} A_n, \qquad \phi^{(n)} = x^n \tag{9-75}$$

Substituting Eqs. (9-75) into Eq. (9-74) gives

$$S_j = A_i \frac{\partial x^i}{\partial x^j}$$

$$= A_i \delta^i_j = A_j \tag{9-76}$$

which demonstrates the equality of S_j and A_j. Thus to prove that $A_{j;k}$ is a tensor for any covariant vector A_j, it is sufficient to prove that this is true for S_j. However, from the form of Eq. (9-73) and since Eq. (9-74) is linear, it is sufficient to treat only the case

$$A_j = \psi \frac{\partial\phi}{\partial x^j} \tag{9-77}$$

Now the quantity in parentheses in Eq. (9-71) multiplied by ψ

$$\psi \frac{\partial^2\phi}{\partial x^j \partial x^k} - \Gamma^i_{jk}\psi\frac{\partial\phi}{\partial x^i} \tag{9-78}$$

is a tensor. Also,

$$\frac{\partial\psi}{\partial x^j}\frac{\partial\phi}{\partial x^k} \tag{9-79}$$

is the outer product of two covariant vectors and, so, is a tensor. Addition of Eqs. (9-78) and (9-79) results in the tensor

$$\frac{\partial}{\partial x^k}\left(\psi\,\frac{\partial\phi}{\partial x^j}\right) - \Gamma^i_{jk}\left(\psi\,\frac{\partial\phi}{\partial x^i}\right) \tag{9-80}$$

Comparison of Eq. (9-80) with Eq. (9-73) completes the proof for the vector $\psi(\partial\phi/\partial x^j)$ and, hence, for any vector A_j.

The concept of a covariant derivative of a vector is easily generalized to treat a covariant tensor of any order.* We consider only type $(0,2)$ tensors explicitly, but this example demonstrates the general rule.

As illustrated in Section 2, not every type $(0,2)$ tensor may be represented by an outer product of two contravariant vectors A_i and B_j. It is possible, however, to represent any type $(0,2)$ tensor by a sum of tensors of the form A_iB_j [1, p. 135]. Thus it is necessary only to derive the covariant derivative of this special case. From Eq. (9-73),

$$A_{i,k} - \Gamma^l_{ik}A_l \tag{9-81}$$

and

$$B_{j,k} - \Gamma^l_{ik}B_l \tag{9-82}$$

are covariant tensors of order 2. Outer multiplication of Eq. (9-81) on the right by B_j and Eq. (9-82) on the left by A_i produces, in both cases, a type $(0,2)$ tensor. Addition of these tensors gives

$$A_{ij;k} = A_{ij,k} - \Gamma^l_{ik}A_{lj} - \Gamma^l_{jk}A_{il} \tag{9-83}$$

where $A_{ij} = A_iB_j$. Equation (9-83) represents the covariant derivative of the tensor A_{ij}, not only for the special case for which it was derived, but for any type $(0,2)$ tensor.

As a final remark, we state that the covariant derivative of the fundamental tensor vanishes identically. Substituting g_{ij} into Eq. (9-83) gives

$$g_{ij;k} = g_{ij,k} - \Gamma^l_{ik}g_{lj} - \Gamma^l_{jk}g_{il}$$
$$= g_{ij,k} - g^{jl}\Gamma_{jik}g_{lj} - g^{il}\Gamma_{ijk}g_{il}$$
$$= g_{ij,k} - (\Gamma_{jik} + \Gamma_{ijk}) \tag{9-84}$$

and from Eq. (9-61), the right-hand side of Eq. (9-84) vanishes.

*We note that the covariant derivative of a contravariant tensor may also be formed, with the resultant tensor being mixed. See [6].

8. Riemann–Christoffel Tensor [1]

Consider the covariant vector A_i, whose covariant derivative is

$$A_{i;j} = A_{i,j} - \Gamma^l_{ij} A_l \tag{9-85}$$

Substituting the tensor $A_{j;k}$ into Eq. (9-83) determines the second covariant derivative of A_i,

$$A_{i;j;k} = A_{i;j,k} - \Gamma^m_{ik} A_{m;j} - \Gamma^m_{jk} A_{i;m} \tag{9-86}$$

Substituting Eq. (9-85) into Eq. (9-86) yields

$$\begin{aligned}
A_{i;j;k} &= (A_{i,j} - \Gamma^l_{ij} A_l)_{,k} - \Gamma^m_{ik}(A_{m,j} - \Gamma^l_{mj} A_l) \\
&\quad - \Gamma^m_{jk}(A_{i,m} - \Gamma^l_{im} A_l) \\
&= A_{i,j,k} - \Gamma^m_{ij} A_{m,k} - \Gamma^m_{ik} A_{m,j} - \Gamma^m_{jk} A_{i,m} \\
&\quad - (\Gamma^l_{ij,k} - \Gamma^m_{ik}\Gamma^l_{mj} - \Gamma^m_{jk}\Gamma^l_{im}) A_l
\end{aligned} \tag{9-87}$$

Obviously, in the second term of the last of Eqs. (9-87), the index change $\Gamma^l_{ij} A_{l,k} \to \Gamma^m_{ij} A_{m,k}$ has been made. Now we wish to form the difference $A_{i;j;k} - A_{i;k;j}$. In the last of Eqs. (9-87), $A_{i,j,k} = A_{i,k,j}$; also, the sum of the second and third terms is symmetric under the interchange of j and k; finally, the Christoffel symbol of the second kind is symmetric under the interchange of its lower indices. Keeping these symmetries in mind, then,

$$A_{i;j;k} - A_{i;k;j} = R^l_{ijk} A_l \tag{9-88}$$

where

$$R^l_{ijk} = \Gamma^l_{ik,j} - \Gamma^l_{ij,k} + \Gamma^m_{ik}\Gamma^l_{mj} - \Gamma^m_{ij}\Gamma^l_{mk} \tag{9-89}$$

The type $(1,3)$ tensor R^l_{ijk} is known as the *Riemann-Christoffel tensor*, or the *curvature tensor*. The significance of the latter name is the following. A necessary and sufficient condition that all of the components g_{ij} be constant is that all of the components R^l_{ijk} vanish. (The necessity of this condition is obvious from Eq. (9-89), and a proof of sufficiency is contained in [2, p. 25].) If g_{ij} is the same at every point of the space, however, g_{ij} may be chosen to be diagonal. Thus for a positive definite metric, if the g_{ij} are constant, the space is Euclidean.* Hence in a Riemannian space, $R^l_{ijk} = 0$ for all components is both necessary and sufficient for the space to be Euclidean.

---

*In a space without a positive definite metric, the condition of constant g_{ij} leads to a flat space.

9. Evolution on the Informational Space–Time Manifold

With this introduction to Riemannian geometry, we now explore its application to molecular genetics. Following the presentation of Section 1 and Chapter 8, Section 3.C, the informational space–time manifold in the x coordinatization is treated as the space $\mathbb{R}^{65}$ composed of n-tuples of the set of real variables $\{x^0, x^1, \ldots, x^{64}\}$. Let $x^0 \equiv t$ stand for the evolutionary time variable, and let the variables x^i, $i = 1, 2, \ldots, 64$, represent the number of codons of type $\alpha_i \in C$ occurring in a physically realizable or physically unrealizable DNA molecule.

Consider the point P_1 in the informational space–time manifold. In light of the previous statements, P_1 represents a DNA molecule at some fixed evolutionary time. Some other point P_2 represents, in general, a different DNA molecule at a different fixed evolutionary time. The motion from P_1 to P_2, then, represents the changing of one DNA molecule into another (i.e., it is a realization of biological evolution). The distance between P_1 and P_2 is the arc length of the curve connecting these two points. The metric is assumed to be positive definite and, hence, the informational space–time manifold to be Riemannian. This assumption is necessary because the concept of a curve of zero length has no known biological counterpart.

The curves in the informational space–time manifold, then, represent the evolutionary progress of DNA molecules. We term these curves *evolutionary motions* and make the following biological postulate:

> *evolutionary motions in the informational space–time manifold are geodesics.*

This postulate serves to select certain curves, the geodesics, and designates these as being entirely descriptive of evolution. Hence the evolutionary motions satisfy the 65 ordinary differential equations [see Eq. (9-59)] ($\mu, v, \sigma = 0, 1, \ldots, 64$)

$$\frac{d^2 x^\mu}{ds^2} = - \Gamma^\mu_{v\sigma} \frac{dx^v}{ds} \frac{dx^\sigma}{ds} \tag{9-90}$$

where s is the arc length of the evolutionary motion. The components $\Gamma^\mu_{v\sigma}$ are determinative of the nature of the evolutionary motion and, hence, may be thought of as comprising an evolutionary *field* on the informational space–time manifold, that is, the $\Gamma^\mu_{v\sigma}$ represent a *biological* evolutionary *force* which determines the path of evolution. In this sense, then, Eqs. (9-90) are evolutionary *equations of motion*. Thus evolution is determined at the molecular level by the curvature of the informational space–time manifold.

The structure of this manifold will be investigated in detail in Chapter 11, and the problem of information loss inherent in the Euclidean vector space formalism (see Chapter 8, Section 3.B) will be addressed as well.

It is significant to notice that when the biological postulate is made, the evolutionary force is totally determined by the *intrinsic* structure of the manifold. The correctness of this view may only be ascertained by an appeal to empiricism (see Chapter 11). However, it is possible that certain evolutionary motions in the informational space–time manifold may exist which are *not* geodesics. In such a case, then, the equations of motion become [2, p. 60]

$$\frac{d^2x^\mu}{ds^2} + \Gamma^\mu_{\nu\sigma}\frac{dx^\nu}{ds}\frac{dx^\sigma}{ds} = z^\mu \tag{9-91}$$

where z^μ is a contravariant vector. This leads to the concept of an *extrinsic* evolutionary force which would operate in conjunction with the *intrinsic* evolutionary force (namely, that arising from the curvature of the manifold) to determine the evolutionary changes in a DNA molecule. Clearly, the existence of extrinsic evolutionary forces and the form of the intrinsic evolutionary force can only be determined by actual comparisons of DNA sequences (see Chapter 11). Consequently we make the simplifying assumption that evolution is determined totally by the intrinsic evolutionary force.

References

1. A. Einstein, *Ann. Physik*, **49**, 769 (1916); reprinted in A. Einstein, H. A. Lorentz, H. Weyl, and H. Minkowski, *The Principle of Relativity*, Dover, New York, 1952, pp. 109–164. All page references to this work pertain to the Dover reprint.

2. L. P. Eisenhart, *Riemannian Geometry*, Princeton University Press, Princeton, NJ, 1926.

3. For a proof of this statement, see T. Levi-Civita, *The Absolute Differential Calculus*, Dover, New York, 1977, p. 5.

4. W. M. Boothby, *An Introduction to Differentiable Manifolds and Riemannian Geometry*, Academic Press, New York, 1975.

5. B. Riemann, *Über die Hypothesen, welche der Geometrie zu Grunde liegen*, H. Weyl, Ed., Springer, Berlin, 1919.

6. P. A. M. Dirac, *General Theory of Relativity*, Wiley, New York, 1975, p. 19.

10 Macromolecular Evolution

1. Evolutionary Theory Meets Molecular Genetics

Evolution describes the dynamic processes through which alterations or mutations in the genetic material of an individual are conveyed throughout an entire population. The mutations themselves occur during some stage in the transmission of genetic information. Indeed, if DNA replication, transcription, and mRNA translation into protein always followed the exact same course, there would be no evolution and the present-day diversification evidenced by the multitude of distinct, disparate species would not have occurred.

Mutations may result in changes in one or a few nucleotide bases in a gene (i.e., point mutations). Alternatively, the overall chromosome number or the spatial arrangement of entire genes on these chromosomes may be affected (i.e., chromosomal mutations). Point mutations result from either the substitution, the addition, or the deletion of a single nucleotide base (or a limited number of nucleotide bases) from the genetic message. Considering the degeneracy of the genetic code, nucleotide substitutions may or may not be reflected in the alteration of the amino acid sequence of a resultant polypeptide. For example, the DNA triplet TTA, which corresponds to the mRNA codon AAU, specifies Asn. When the third nucleotide position is changed to G (C in mRNA), the triplet still codes for Asn. If, however, the third position is changed instead to T in DNA (A in mRNA), Lys will be inserted into the protein in place of Asn. The replacement of one amino acid for another as a result of a substitution in nucleotide base sequence is termed a missense mutation. Such mutations may have little overall effect on the system if they appear along a relatively noncritical region of the protein. However, if the protein is an enzyme and the change occurs at the active site, the catalytic ability of the protein may be adversely affected. Substitution errors can also produce the exchange of a codon specifying one of the 20 amino acids for one that codes for termination [e.g., UGG($=$Trp)$\rightarrow$UGA($=$TC)]. These so-called nonsense mutations result in an abbreviated polypeptide chain. Additions or deletions to the nucleotide base sequence produce a different reading (i.e., frame-shift mutations). As a result, the rest of the genetic message, downstream from the point of mutation, will be interpreted in a garbled manner, evidenced by a new amino acid sequence. In the event that the addition or deletion involves three nucleotide bases (or a multiple of three nucleotide bases), the resultant protein will display an extra amino acid or the loss of one, respectively. The remainder of the protein will be entirely normal.

Point mutations occur spontaneously in nature with a characteristic frequency. However, the mutation rate can be increased considerably in the presence of ionizing radiation or the addition of chemical mutagens.* The frequency of spontaneous point mutations varies from organism to organism as well as within different genes of the same individual. Common mutation rates range from 10^{-9} to 10^{-6} per gene per cell division in prokaryotes to 10^{-6} to 10^{-4} per gene per gamete in eukaryotes [1]. Although the mutation rate for a single organism is quite small, because every individual has a number of genes and species consist of many organisms, the overall effect of such mutations is significant.

In this context, when speaking of evolutionary changes it is best to speak in terms of their effects on populations and not individuals. Clearly, for a mutation to be significant in an evolutionary sense, the individual possessing the mutation must pass this change to its progeny. Thus the genetic alteration of a population is facilitated by mutation, migration, random drift, and natural selection. Of course, mutation of the genetic material is the fundamental source of any variation observed at the macroscopic level. Migration, or gene flow, describes the distribution of genetic variation first within a population, and then between populations of the same species. Random drift is essentially an error in genetic frequency introduced as a consequence of sampling a finite population. As such, it represents a more significant consideration when dealing with small populations.

Natural selection is the process by which an alternative genetic message possessing an adaptive advantage is reproduced preferentially (at the expense of a less adaptive one) and spreads throughout the population. This is reasonable since an organism with such a selective advantage will be more successful and produce more offspring, thereby increasing the frequency of this adaptive variation in the population. Clearly, natural selection can only occur in the presence of genetic variation. Thus such variation is determinative of the evolutionary potential of a species. (For more details see [2–6].)

Considerable variation, as evidenced by genetic polymorphism or hetero-zygosity, is found throughout all natural populations [7]. In a natural population, at least 30–80% of all structural genes are polymorphic. In addition, approximately 5–20% of an individual's structural genes are heterozygous. However, the mutation of regulatory genes may play a more

---

*A mutagen is a chemical that increases the frequency of point mutations. Examples include the acridines and various alkylating agents.

significant role in determining evolution than do changes in structural genes [8]. Alterations in regulatory genes could effectively influence the expression of structural genes without necessarily changing the sequence of these genes themselves.

Following the determination of genetic polymorphism, two diametrically opposed theories were developed to account for this observation. According to the neutralist theory, reported variations merely represented an artifact of population size differences [9–11]. Accordingly, the genetic variation evidenced in natural populations is biologically and adaptively insignificant because it is neither useful nor harmful, just neutral. Thus polymorphism is nothing more than a kind of "noise" in the genetic system [11]. This viewpoint was later modified as the mutation–equilibrium theory, which states that random fixation of neutral or nearly neutral mutations is the primary cause of molecular evolution [12–15]. In this case, there is thought to exist a continuum of mutations, ranging from neutral ones (possessing no advantage or disadvantage for the organism) to biologically significant ones, upon which selection can build. Each individual mutation achieves a certain intermediate genetic variation frequency depending on whether they constitute selectively neutral, positive, or negative variants. Evolution then is determined by the equilibrium between mutation and selection, where the latter is given only a minor deterministic role in the process. Conversely, the opposite camp maintains a selection or adaptive strategy theory of evolution [4, 6, 16, 17]. Here polymorphism is viewed as the raw material upon which selection acts in an important way. This theory invokes frequency-dependent and other types of balancing selection and views directional selection as the major cause of allele fixation. Thus a selectionist is primarily interested in explaining how the observed genetic variation affects the "fitness" of a population.

Evolution may be thought of as proceeding through two distinct phases: anagenesis and cladogenesis [18]. Anagenesis or phyletic evolution monitors the time course of adaptations occurring within a given phylogenetic line (e.g., the evolution of humans). Cladogenesis describes the process by which a single phylogenetic line diverges or splits into two or more independently evolving lines. The final and decisive step in a cladogenetic procedure is the formation of two distinct (i.e., separately evolving) species.

Species consist of groups of populations that are effectively isolated from other such groups and thereby constitute a separate evolutionary unit. Anagenic changes which transpire in individual organisms of a local population can spread to other individuals in the same population and to other populations of the same species by natural selection, but they may not be passed to another species.

Whereas a direct comparison with ancestral forms is often impossible, anagenic progress within a species may be monitored by comparing the species of interest with other living closely related species. As an example, the degree of similarity in molecular composition among living species may be utilized to determine their phylogenetic relationship to one another. In this context, gel electrophoresis, immunological antibody–antigen assays, amino acid sequencing of proteins, and other techniques have provided a significant contribution to the construction of phylogenetic history. In the past the structural variation of DNA was followed indirectly by examining the end product of expression (i.e., amino acid sequences of proteins, see below). However, with the advent of rapid sequencing techniques and DNA hybridization technology, the direct study of DNA base sequence homology is also possible.

2. Construction of Evolutionary Metrics and Diagrams

The amino acid sequences of protein polypeptide chains are images of segments of the genome and reflect the fine structure of the corresponding pieces of DNA. Therefore the structure of proteins may be considered the current molecular end products of the evolutionary variations experienced by their genes. Crick [19] was one of the first to recognize the significance of this relationship when he stated:

> Biologists should realize that before long we shall have a subject which might be called "protein taxonomy," the study of amino acid sequences of proteins of an organism and the comparison of them between species. It can be argued that these sequences are the most delicate expression possible of the phenotype of an organism and that vast amounts of evolutionary information may be hidden away within them.

Due to the amount of work involved in obtaining numerous amino acid sequences and the difficulty of gathering large enough samples of pure proteins from a diversity of species, the data collection has progressed slowly. However, compendia of such sequences exist [20].

Taxonomy or phylogeny studied through the structures of informational macromolecules offers several advantages over organism-level morphological taxonomy. Proteins represent the most subtle phenotypic expression. In addition, the chemistry of these proteins is related in a direct way to the fine structure of their genes. Changes in proteins are the expression of fundamental evolutionary unit changes, that is, the evolutionary fixation of single nucleotide substitutions in the corresponding segments of the genome. Definite constraints are placed on protein evolutionary variations which, though necessarily present, cannot possibly be deciphered in morphological evolutionary alterations represented by complicated changes which often encompass numerous genes [21]. Protein evolutionary changes are, in an observable fashion, restricted by both the genetic code and the phenotypic requirement of maintaining a functional structure. Therefore one would expect that a thorough understanding of the structure–function relation of a protein, coupled with the information extracted from the statistical techniques of homologous protein amino acid sequence analyses [22, 23], may lead to a more satisfying view of evolution. Such a description would include a detailed reconstruction of the complete evolutionary tree of the protein, a delineation of its evolutionary relations to other gene products, and quite possibly an understanding of the rules and regulations that govern molecular evolution [21].

Cytochromes of the c type have achieved ubiquitous biological distribution in the mitochondrial respiratory chain of eukaryotes, in the oxidative systems of many prokaryotes, and in the photosynthetic membranes of both eukaryotes and prokaryotes, including those that do not use molecular oxygen as a terminal electron acceptor. Since the first determination of the primary structure of horse cytochrome c [24], the amino acid sequences of cytochrome c from a wide taxonomic range of species have appeared. All cytochrome c species studied are small proteins with slightly over 100 residues. Cytochromes c of the mitochondrial or eukaryotic type are strongly basic, have extensive similarities of amino acid sequence, and react identically in cytochrome c depleted mitochondria from mammalian species [25] and with cytochrome c oxidase and reductase preparations [26]. Bacterial cytochrome c is more variable in nature than that of eukaryotes. Some are acidic and others strongly basic, such as cytochrome c_3 of sulfate-reducing organisms. Since bacterial c-type cytochromes were known to be unreactive with mammalian cytochrome c oxidase preparations, they were formally considered to be unrelated to the eukaryotic cytochromes c. However, as the amino acid sequences of several bacterial cytochromes c were determined [27, 28], obvious similarities between their structures and those of eukaryotic cytochromes c became apparent. Thus it appears that c-type cytochromes of eukaryotes, prokaryotes, and photosynthetic membranes are phylogenetically related, or homologous, proteins. Cytochromes c, therefore, may be considered proteins of choice, not only to examine the phylogenetic relationship of various groups of prokaryotes, but also to determine unambiguously the phylogenetic relations between prokaryotes and eukaryotes. As such, their study might possibly provide the necessary insight required to unravel the history of that most important of evolutionary transitions, namely, the organization of intracellular organelles, events that led to the enormous diversity of life presently in existence [21].

As the amino acid sequences of the eukaryotic cytochromes c became available, it was evident that the primary structure of this protein could accommodate considerable variation, and that the degree of variability was related to the phylogenetic distance between the species carrying the protein [29]. In addition, it was observed that the cytochromes c of mammals were equally different from the cytochromes c of birds, that the proteins of mammals and birds considered together were equally different from those of fish, that those of all vertebrates were equally different from those of insects, and that the cytochromes c of all animals, vertebrates, and invertebrates were roughly equally different from those of fungi [29]. Therefore the time elapsed since evolutionary divergence appears to be the parameter determining by how many residues the cytochromes c of two different species vary, implying that residue changes in the protein were fixed in the course of evolution at a roughly constant rate [29]. Using as a standard the known time of divergence of mammalian and avian lines of descent (280 million years ago), it is possible to define a *unit evolutionary period* for cytochrome c, that is, the evolutionary time required to produce a single change in the cytochromes c of two diverging lines of evolutionary descent [30]. This period was calculated to be 26.4 million years for amino acid residue substitution and 21.4 million years for the nucleotide changes in the corresponding codons.

The amino acid sequences of several eukaryotic cytochromes c indicate that 26 residues have remained constant, and that 75% of the molecule has been subject to amino acid substitution [21]. Amino acid substitution may be divided into two types:

1. Conservative substitutions, for which the alternative amino acids are so similar in structure or physicochemical properties that they may be capable of identical functions. Examples of this type would include the exchange of two strongly basic amino acids (lysine ↔ arginine) or substitutions involving only hydrophobic amino acids (valine ↔ leucine ↔ isoleucine ↔ phenylalanine ↔ tyrosine).

2. Radical substitutions. Other locations along cytochrome c, however, are subject to radical substitution between amino acids that are so dissimilar that they appear unlikely to be importantly involved in maintaining the tertiary structure or the function of the protein.

The existence of invariant positions, which undergo no substitution from species to species, indicates that the residues involved at these points have a function that is compatible only with their particular structures. Since this invariance spans broad taxonomic lines, in the case of cytochrome c, this function must be of an important general nature and not one that is adapted only to a particular species.

The next step in the study of the molecular variation of homologous protein molecules is to construct an approximation for the nature of their relationship through a phylogenetic tree. A statistical procedure for obtaining such a history from amino acid sequences was developed by Fitch and Margoliash [22]. In order for the derivation of the similarity relations for a set of amino acid sequences to have any biological significance, two conditions must be satisfied: (1) the structures must have more than a random degree of similarity, and (2) the proteins must be truly homologous, that is, all descended from a common ancestral form. To look for significant similarities between amino acid sequences, all possible segments of a definite length from the chains in question must be compared [31]. The minimal mutation or replacement distance is computed as the minimal number of single nucleotide substitutions required to transform the gene segment coding for one protein into that coding for the other. When replacement distances are plotted against the number of times each occurs for all comparisons between human and baker's yeast iso-1 cytochrome *c*, Figure 10-1 results. The comparisons indicating random similarity fall within the Gaussian area of the distribution curve. The highest point of this portion of the curve corresponds to a replacement distance of 45, as expected, since 30-residue segments were compared in this analysis, and on the average, it takes about 1.5 mutations to transform any codon to any other codon according to the genetic code [31a]. The portion of the distribution curve to the left of the peak consists of those comparisons for which the replacement distances are smaller than expected for purely random similarities. Thus its presence indicates the existence of a significant degree of

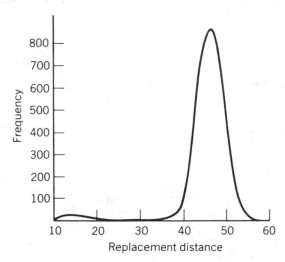

Figure 10-1. Comparison of minimal replacement distance for all possible 20-residue segments of human and baker's yeast cytochrome *c*. (*After* [*31b*].)

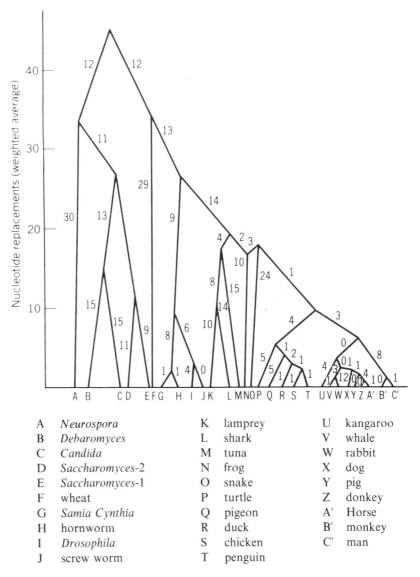

A	Neurospora	K	lamprey	U	kangaroo
B	Debaromyces	L	shark	V	whale
C	Candida	M	tuna	W	rabbit
D	Saccharomyces-2	N	frog	X	dog
E	Saccharomyces-1	O	snake	Y	pig
F	wheat	P	turtle	Z	donkey
G	Samia Cynthia	Q	pigeon	A'	Horse
H	hornworm	R	duck	B'	monkey
I	Drosophila	S	chicken	C'	man
J	screw worm	T	penguin		

Figure 10-2. Statistical phylogenetic tree based on the minimal replacement distances between the cytochromes c of the species listed. (*After [22]*.)

similarity between human and baker's yeast cytochromes c, that is, they are homologous proteins.

A word of caution must also be included in this discussion. Lack of significant similarity from this test does not necessarily mean that the proteins did not at one time have a common ancestral form. It only indicates that if they are indeed homologous, the changes they have incurred since their divergence are so large as to make it impossible to see any evidence of this common ancestry today. Conversely, the presence of significant similarities may be presumed evidence for divergence from a common source or may be the result of the functional convergence of independent evolutionary stocks. This impass can be resolved if one can read the structures of these proteins at two distinct times in the history of their evolutionary descent. To accomplish this, putative ancestral sequences must be constructed.

Figure 10-2 depicts a phylogenetic tree for several different cytochromes c [22].* This representation is based solely on amino acid sequences and the genetic code, where the minimal replacement distances are calculated for all possible comparisons of the amino acid sequences [for n sequences there are $n(n-1)/2$ replacement distances]. One begins by joining together the two proteins that show the smallest replacement distance, and calculates the average replacement distances of all other proteins with respect to these first two, now considered as a single subset. The next protein to be joined to the tree is the one with the smallest replacement distance in comparison to the first two. This procedure is repeated until all proteins have been linked. The number of nucleotide substitutions required between various points on the tree can be calculated and one can determine the replacement distances between any two proteins by summing over the appropriate segments. The phylogenetic tree presented for cytochrome c is a reasonable representation of the phylogeny of the species displayed with some reservations. For example, the primates branch off the ancestral mammalian line before the marsupials, the turtles appear to be more closely related to the birds than to the rattlesnake, and the shark is closer to the lamprey than to the tuna. As the amino acid sequences of more species are included in this scheme, these inconsistencies should be eliminated.

The statistically derived phylogenetic tree can be used to determine the amino acid sequences of ancestral proteins at each branch point. In the original method of Fitch and Margoliash [22], this was done by choosing an amino acid residue in an ancestral sequence for which, during its phylogenetic descent, its corresponding codon required the least number of mutations, the fewest segments of the phylogenetic tree containing multiple variations, and so

---

*The procedure used by Fitch and Margoliash to determine distances along the phylogenetic tree has been questioned since their metric it does not satisfy the triangle inequality. In this connection, see [32].

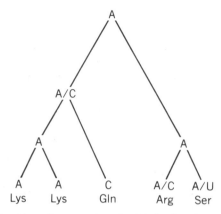

Figure 10-3. Reconstruction of ancestral nucleotide for first base of amino acid codons indicated, given a specific topology of protein relationships. (*After* [*21*].)

on. The idea was to arrive at an approximation to actual evolutionary transitions. From the phylogenetic tree for cytochrome *c* it is possible to show that the ancestral primate cytochrome *c* differed from the ancestral mammalian protein at five residue positions, while the line of descent of human cytochrome *c* differed from the monkey protein line by a single mutation, which occurred in the human and not in the monkey line of descent [22].

Fitch [31b] later revised the procedure to one in which the descent of each nucleotide in each codon of the amino acid sequence is reconstructed separately. A diagram of this procedure is provided in Figure 10-3. The nucleotide assigned to each coding position at the top is one that is common to both immediate descendents (the intersection of the descendent nucleotide sets), or when the intersection results in an empty sct (i.e., no common nucleotide among the descendents) then the top position is given to all nucleotides in the two immediate descendents (the union of the descendent sets) [21]. At a second stage, a nucleotide replacement is noted whenever an intersection would have yielded an empty set, so that the minimal number of such mutations are required to account for the total phylogeny at the particular coding position under study. When completed, this procedure may be used to reconstruct the putative amino acid sequences of ancestral proteins [21].

The technique of ancestral amino acid sequence reconstruction can distinguish between analogous (but convergent) and homologous (but divergent) sets of proteins. If the ancestral forms are the same but the descendents are different, evolutionary divergence has occurred. Conversely, if the ancestral forms are different and similarities are developed only among the descendents, then convergence has transpired. This procedure, therefore, presents for the first time a statistical demonstration that all eukaryotic cytochromes c are descendents of a common ancestral form which had experienced evolutionary divergence and thereby provides support for the unitary theory of the origin of life.

The study of molecular variations in homologous proteins, as developed by Fitch and Margoliash, has fostered numerous related investigations. The work on cytochrome c has been extended to include more species, and the technique has been applied to other groups of homologous proteins. Clearly, however, the most exciting prospect for such homology studies has come from its more recent extension to the realm of DNA nucleotide base sequences (see [33] and reference, therein). Such studies will provide the most fundamental information yet available concerning the mutations or variations that contribute to the evolution of a species.

References

1. F. J. Ayala, in *Molecular Evolution*, F. J. Ayala, Ed., Sinauer Associates, Sunderland, MA, 1976.
2. T. Dobzhansky, *Genetics of the Evolutionary Process*, Columbia University Press, New York, 1970.
3. E. Mayr, *Population, Species and Evolution*, Belknap, Cambridge, MA, 1970.
4. F. J. Ayala, Ed., *Molecular Evolution*, Sinauer Associates, Sunderland, MA, 1976.
5. T. Dobzhansky, F. J. Ayala, G. L. Stebbins, and J. W. Valentine, *Evolution*, Freeman, San Francisco, CA, 1977.
6. R. Milkman, Ed., *Perspectives on Evolution*, Sinauer Associates, Sunderland, MA, 1982.
7. R. C. Lewontin, *The Genetic Basis of Evolutionary Change*, Columbia University Press, New York, 1974.
8. A. C. Wilson, in *Molecular Evolution*, F. J. Ayala, Ed., Sinauer Associates, Sunderland, MA, 1976.
9. M. Kimura, *Nature*, **217**, 624 (1968).
10. M. Kimura, *Proc. Natl. Acad. Sci. USA*, **63**, 1181 (1969).
11. M. Kimura and T. Ohta, *Theoretical Aspects of Population Genetics*, Princeton University Press, Princeton, NJ, 1971.
12. T. Ohta and M. Kimura, *Genetics*, **76**, 615 (1974).
13. T. Ohta and M. Kimura, *Amer. Natur.*, **109**, 137 (1975).
14. M. Kimura, *Nature*, **267**, 275 (1977).
15. M. Kimura, *Proc. Natl. Acad. Sci. USA*, **78**, 5773 (1981).
16. R. Levins, *Evolution in Changing Environments*, Princeton University Press, Princeton, NJ, 1968.

17.   R. K. Selander and D. W. Kaufman, *Proc. Natl. Acad. Sci. USA*, **70**, 1875 (1973).

18.   B. Rensch, *Evolution above the Species Level*, Columbia University Press, New York, 1959.

19.   F. H. C. Crick, *Symp. Soc. Exp. Biol.*, **12**, 138 (1958).

20.   M. O. Dayhoff, *Atlas of Protein Sequences and Structures*, Natl. Biomedical Research Foundation, Washington, DC, 1972, and further supplements.

21.   E. Margoliash, *Adv. Chem. Phys.*, **29**, 191 (1975).

22.   W. M. Fitch and E. Margoliash, *Science*, **155**, 279 (1967).

23.   W. M. Fitch and E. Margoliash, in *Evolutionary Biology*, vol. 4, T. Dobzhansky, M. K. Hecht, and W. C. Steere, Eds., Plenum, New York, 1970.

24.   E. Margoliash, E. L. Smith, G. Kreil, and H. Tuppy, *Nature*, **192**, 1125 (1961).

25.   V. Byers, D. Lambeth, H. A. Hardy, and E. Margoliash, *Fed. Proc.*, **30**, 1286 (1971).

26.   L. Smith, M. E. Nava, and E. Margoliash, in *Oxidases and Related Redox Systems*, T. E. King, H. S. Mason, and M. Morrison, Eds., University Park Press, Baltimore, MD, 1973.

27.   M. D. Kamen and T. Horio, *Ann. Rev. Biochem.*, **39**, 673 (1970).

28.   R. P. Ambler, M. Bruschi, and J. LeGall, in *Recent Advances in Microbiology*, A. Perez-Miravete and D. Palaez, Eds., Asoc. Mexicana de Micro., Mexico City, 1971.

29.   E. Margoliash, *Proc. Natl. Acad. Sci. USA*, **50**, 672 (1963).

30.   E. Margoliash and E. L. Smith, in *Evolving Genes and Proteins*, V. Bryson and H. J. Vogel, Eds., Academic Press, New York, 1965.

31.   This procedure was developed by W. M. Fitch, in (a) *J. Mol. Biol.*, **16**, 9 (1966); (b) *J. Mol. Biol.*, **49**, 1 (1970).

32.   W. A. Beyer, M. L. Stein, T. F. Smith, and S. A. Ulam, *Math. Biosci.*, **19**, 9 (1974).

33.   R. K. Selander, in *Perspectives on Evolution*, R. Milkman, Ed., Sinauer Associates, Sunderland, MA, 1982.

11 Realization of Molecular Genetics as a Differential Geometry

1. Introduction

At the molecular level, the *information content** of a gene is dictated by the linear arrangement of the nucleic acid bases in a DNA molecule. Perhaps one of the most exciting prospects in molecular genetics is the possibility of investigating biological evolution through changes in DNA base sequences. Such analyses have not been practicable formerly because of the lack of detailed DNA sequence data. Analogous studies at the protein level, on the other hand, have benefited from the availability of protein sequence data [1] and are well established (see [2–4] and Chapter 10, Section 2). However, with the advent of rapid DNA sequencing techniques [5] it is certain that compendia of DNA sequences, from various parts of the genome in diverse organisms, will more and more become available.

In this chapter we offer the details and qualitative implications of a mathematical formalism in which changes in the information content of a DNA molecule undergoing evolution may be described [6, 7]. The present theory is intended to correct the inherent shortcomings of our previous attempts to interpret molecular genetics in a linear space (see Chapter 8). The current approach, which was briefly introduced in Chapter 9, Section 9, is founded on a 65-dimensional differentiable manifold (the *informational space–time manifold*) in a coordinatization such that the manifold points represent (1) the number of each codon type in a DNA molecule, and (2) the evolutionary time of that DNA.

In order to further develop the ideas outlined in Chapter 9 (Section 9), Section 2 of this chapter explores the tenet that curves in the informational space–time manifold may be interpreted as representing the evolutionary progress of DNA molecules. A provisional postulate is discussed in which evolutionary motions in this manifold are described as geodesics. It is

---

*Our use of the term "information content" is consonant with the terminology of molecular genetics in which DNA, RNA, and protein are referred to as "informational macromolecules." However, by this usage we intend no reference to information theory.

subsequently shown that the intrinsic structure of the manifold determines a biological *evolutionary field*, and *evolutionary equations of motion* are elaborated. The essential result is that the solution to evolutionary questions which are formulated at the DNA level resides, in principle, in the intrinsic structure of the informational space–time manifold, that is, in the knowledge of the biologically correct *genetic cosmology*.

The intrinsic structure of the manifold is, of course, determined by the fundamental tensor (see Section 2), and in Section 3 we investigate a genetic cosmology in which the fundamental tensor is diagonal and a function only of the evolutionary time. Finally, the evolutionary equations of motion for a weak field, which is solely evolutionary time dependent, are given and the nature of empirical input into genetic cosmology is discussed.

2. Evolutionary Motions Treated as Geodesics

A. Geodesics

In terms of local differential geometry (see [8] and Chapter 9), the differential element of length ds in the informational space–time manifold M in the x coordinatization, $M[x]$, is given by*

$$ds^2 = g_{\mu\nu}dx^\mu dx^\nu \tag{11-1}$$

where $g_{\mu\nu} = g_{\mu\nu}(x^0, x^1, \ldots, x^{64})$ and $g_{\mu\nu}$ is a symmetric, covariant, order-2 tensor having nonzero determinant. The invariant ds^2 is termed the *metric*, and a differentiable manifold having a positive definite metric is said to be *Riemannian*. The tensor $g_{\mu\nu}$ is referred to as the *fundamental tensor* and is totally determinative of the intrinsic structure of M (see below).

Consider two points M_1 and M_2 in $M[x]$ and a curve C connecting these two points. If the coordinates of the points of C are given as functions of a general parameter τ, such that $x^\mu(\tau_1)$ and $x^\mu(\tau_2)$ are the coordinates of M_1 and M_2, respectively, then we may define the integral

$$s = \int_{\tau_1}^{\tau_2} \left[g_{\mu\nu} \frac{dx^\mu}{d\tau} \frac{dx^\nu}{d\tau} \right]^{1/2} d\tau \tag{11-2}$$

where s is the arc length of the curve C between points M_1 and M_2. If s is stationary (that is, if, holding the endpoints of C constant, the first-order variation of s vanishes), then C is a *geodesic*, which is simply the generalization

---

*We adopt the Einstein summation convention (see Chapter 9, Section 1).

of the Euclidean straight line to curved spaces. By the techniques of Chapter 9, the stationarity of C implies

$$\frac{d^2 x^\mu}{ds^2} + \Gamma^\mu_{v\sigma} \frac{dx^v}{ds} \frac{dx^\sigma}{ds} = 0 \qquad (11\text{-}3)$$

where the parameter τ has been taken to be the arc length s, and where the *Christoffel symbols* $\Gamma^\mu_{v\sigma}$ are defined as

$$\Gamma^\mu_{v\sigma} = g^{\mu\gamma} \Gamma_{\gamma v\sigma} \qquad (11\text{-}4)$$

and

$$\Gamma_{\gamma v\sigma} = \tfrac{1}{2}(g_{\gamma v,\sigma} + g_{\gamma\sigma,v} + g_{v\sigma,\gamma}) \qquad (11\text{-}5)$$

In Eq. (11-5) we have used the notation

$$g_{\gamma v,\sigma} = \frac{\partial g_{\gamma v}}{\partial x^\sigma} \qquad (11\text{-}6)$$

The extremals of Eq. (11-2), where the parameter τ is taken to be the arc length s, are the integral curves of the 65 ordinary differential equations given in Eq. (11-3). These are the geodesic curves, and they satisfy the condition that anywhere along the curve (cf. Chapter 9)

$$g_{\mu v} = \frac{dx^\mu}{ds} \frac{dx^v}{ds} = 1 \qquad (11\text{-}7)$$

B. Evolutionary Equations of Motion

In a manifold that does not possess a positive definite metric, there exist curves of zero length. Since such a curve in $M[x]$ appears, at present, to have no biological meaning, we take $M[x]$ to be Riemannian. In addition, we make the following postulate:

> *evolutionary motions in the informational space–time manifold are geodesics.*

This postulate selects certain curves, the geodesics, and discriminates these as being entirely descriptive of evolution. Hence the evolutionary motions satisfy Eq. (11-3). The components $\Gamma^\mu_{v\sigma}$ are determinative of the nature of an evolutionary motion and, therefore, may be thought of as comprising an *evolutionary field* on the informational space–time manifold. In this sense, then, Eq. (11-3) represents evolutionary equations of motion. Since the $\Gamma^{\mu}_{\nu\sigma}$ determine the curvature of the manifold [8], it follows that evolution may be viewed as resulting from the curvature of the informational space–time manifold.

The identification of geodesics with evolutionary motions is provisional. In fact, this postulate may be generalized to include the case in which not all of evolution is governed by the curvature of the informational space–time manifold [9]. Such a generalization requires, in a straightforward manner, the introduction of the concept of an *extrinsic* evolutionary field.

3. Genetic Cosmology

The fundamental result of the previous section is that the solution to evolutionary questions formulated at the DNA level resides, in principle, in the intrinsic structure of the informational space–time manifold, that is, in the biologically correct *genetic cosmology*.

In examining the manifold structure, we discuss three cases: (1) the absence of an evolutionary field, (2) the presence of a permutational evolutionary field, and (3) an (incomplete) model genetic cosmology in which the evolutionary field is a function only of the evolutionary time. A comparison of the results of (1) and (2) resolves the information loss discussed in Chapter 8. Finally, we comment briefly on empirical input into genetic cosmology.

A. Absence of an Evolutionary Field

The rectilinear propagation, in evolutionary time, of a DNA molecule which is not evolving is specified by the following two conditions:

C1.   The $g_{\mu\nu}$ are constant for all μ, ν.
C2. $v^i \equiv dx^i/ds = 0$ for all i.

(C1) ensures that the manifold be rectilinear (flat), and since we are considering only Riemannian metrics, the manifold must be Euclidean. The second condition ensures that all of the informational space coordinates be constant along an evolutionary motion.

For an evolutionary motion, application of (C2) to Eq. (11-7) yields

$$g_{00}v^{0^2} = 1 \qquad (11\text{-}8)$$

or

$$v^0 = g_{00}^{-1/2} \qquad (11\text{-}9)$$

Thus the evolutionary equation of motion is

$$\frac{dv^0}{ds} = \frac{d}{ds}(g_{00}^{-1/2}) = 0 \tag{11-10}$$

where the second equality follows from (C1).

Now, from Eq. (11-8) we have

$$ds = g_{00}^{1/2}\, dx^0 = g_{00}^{1/2}\, dt \tag{11-11}$$

Because of (C1), however, we may choose $g_{00} = 1$. The arc length of the evolutionary motion from point M_1 to point M_2 is then given by

$$s = \int_{t_1}^{t_2} dt = \Delta t \tag{11-12}$$

which is, of course, the same result as that of Chapter 8. Thus a DNA molecule which is not evolving is characterized by 64 coordinates x^i, which are constant, and coordinate $x^0 \equiv t$, which is propagating in a linear manner.

B. Presence of a Permutational Evolutionary Field

We next consider a DNA molecule which is evolving such that only the linear order of its codons changes, that is, the informational space coordinates are constant along an evolutionary motion. Such evolution is specified by the following two conditions:

C′1. $g_{\mu v} = \begin{cases} g_{\mu\mu}(t), & \text{for } \mu = v \\ 0, & \text{for } \mu \neq v \end{cases}$

C′2. $v^i = 0,$ for all i

For an evolutionary motion, application of (C′2) to Eq. (11-7) yields

$$g_{00}v^{0^2} = 1 \tag{11-13}$$

Clearly, the difference between Eqs. (11-13) and (11-8) is that g_{00} is a function of evolutionary time in Eq. (11-13), but g_{00} is a constant in Eq. (11-8).

Rewriting Eq. (11-13) as

$$v^0 = g_{00}^{-1/2} \tag{11-14}$$

we find that the evolutionary equation of motion is

$$\frac{dv^0}{ds} = \frac{d}{ds}(g_{00}^{-1/2})$$

(11-15)

However, Eq. (11-13) implies

$$ds = g_{00}^{-1/2}\,dx^0 \equiv g_{00}^{1/2}\,dt$$

(11-16)

and, hence, we find

$$\frac{dv^0}{ds} = g_{00}^{-1/2}\frac{dv^0}{dt}; \qquad \frac{d}{ds}(g_{00}^{-1/2}) = g_{00}^{-1/2}(g_{00}^{-1/2})_{,0}$$

(11-17)

Substituting Eq. (11-17) into Eq. (11-15), the evolutionary equation of motion becomes

$$\frac{dv^0}{dt} = (g_{00}^{-1/2})_{,0}$$

(11-18)

Finally, from Eq. (11-16) the arc length of the evolutionary motion from point M_1 to point M_2 is given by

$$s = \int_{t_1}^{t_2} g_{00}^{1/2}(t)\,dt$$

(11-19)

where we have explicitly indicated the evolutionary time dependence of $g_{00}^{1/2}$.

The content of Eq. (11-19) is elucidated by comparison with Eq. (11-12). For the case in which evolution is manifested through codon permutation only, the arc length of the evolutionary motion between two points in the informational space–time manifold is a *nonlinear* function of the evolutionary time coordinates of the points. In the absence of an evolutionary field, however, the arc length of an evolutionary motion between two points in the informational space–time manifold is a *linear* function of the evolutionary time coordinate of the points.

The choice of a Euclidean manifold structure inherently results in an information loss which identifies the distance between two points in $M[x]$ regardless of whether they represent the same DNA molecule at different evolutionary times or two distinct DNA molecules, varying only in codon order, at different evolutionary times (see Chapter 8). From the above derivation we conclude that a curved informational space–time manifold restores the information loss, with respect to evolution, that is inherent in the original Euclidean formulation.

C.   An (Incomplete) Model Genetic Cosmology

We shall now consider a model genetic cosmology which incorporates the results of Sections 3.A and 3.B as special cases. In addition, we make the assumption that the evolutionary field is weak (i.e., that biological evolution occurs slowly). Thus the conditions are

C"1. $g_{\mu\nu} = \begin{cases} g_{\mu\mu}(t), & \text{for } \mu = \nu \\ 0, & \text{for } \mu \neq \nu \end{cases}$

C"2. $\lim\limits_{t \to \infty} g_{\mu\mu} = c_{\mu\mu}$, where the $c_{\mu\mu}$ are constants for all μ.

C"3. v^i is a small quantity of first order with respect to v^0 for all i.

The model is incomplete in the sense that we never specify an exact functional form for the $g_{\mu\nu}$.

In view of (C"1) we find that the only nonzero $\Gamma^{\mu}_{\nu\sigma}$ are

$$\Gamma^i_{i0} = \tfrac{1}{2}g^{ii}g_{ii,0}, \qquad \Gamma^0_{ii} = -\tfrac{1}{2}g^{00}g_{ii,0}, \qquad \Gamma^0_{00} = \tfrac{1}{2}g^{00}g_{00,0} \qquad (11\text{-}20)$$

Substitution of Eqs. (11-20) into Eq. (11-3) yields the evolutionary equations of motion

$$\frac{dv^i}{ds} = -g^{ii}g_{ii,0}v^i v^0 \qquad (11\text{-}21a)$$

and

$$\frac{dv^0}{ds} = -\frac{1}{2}g^{00}\{g_{00,0}v^{0^2} - g_{ii,0}v^{i^2}\} \qquad (11\text{-}21b)$$

Imposing condition (C"2) we see that the equations of motion become

$$\frac{dv^\mu}{ds} = 0 \qquad (11\text{-}22)$$

in the limit of $t \to \infty$, thus regenerating the results of Section 3.A. For those evolutionary motions for which $v^i = 0$, the equations of motion become*

$$\frac{dv^0}{dt} = (g_{00}^{-1/2})_{,0} \tag{11-23}$$

thus regenerating the results of Section 3.B.

Equations (11-21a, b) may be simplified by applying (C''3). For an evolutionary motion, then, we find [see (C''1) and Eq. (11-7)]

$$g_{00}v^{0^2} = 1 \tag{11-24}$$

where we have neglected terms which are of the second order in smallness. Applying (C''3) to Eqs. (11-21a, b) and making use of Eq. (11-24), we find

$$\frac{dv^i}{dt} = -g^{ii}g_{ii,0}v^l \tag{11-25a}$$

and

$$\frac{dv^0}{dt} = (g_{00}^{-1/2})_{,0} \tag{11-25b}$$

where all second-order terms have been neglected.

Equations (11-25a, b) are the evolutionary equations of motion for a weak evolutionary field which is a function only of the evolutionary time.

We may rewrite Eq. (11-25a) as

$$\frac{dv^i}{dt} = (\ln g^{ii})_{,0}v^i \tag{11-26}$$

Remembering that $v^i \equiv dx^i/ds$ and integrating Eq. (11-26) twice over the interval $[0, t]$, we find

$$x^i(t) = \int_0^t g_{00}^{1/2}\left\{\int_0^{t'} g_{00}^{-1/2}(\ln g^{ii})_{,0}\frac{dx^i}{dt''}\,dt''\right\}dt' + \left[g_{00}^{-1/2}\frac{dx^i}{dt}\right]_{t=0}t + [x^i]_{t=0} \tag{11-27}$$

This equation represents a formal solution to the question of how the codon population of a DNA molecule changes with evolutionary time, in the presence of a weak evolutionary field which is a function only of the evolutionary time.

---

*The reader should note that when $g_{\mu\nu} = 0$ for all $\mu \neq \nu$, it follows that $g^{\mu\mu} = g_{\mu\mu}^{-1}$.

Further investigations of Eq. (11-27) require a choice of the time functionality of the fundamental tensor components $g_{\mu\mu}$. Only then does the theory
presented here become a complete model genetic cosmology.

D. Empirical Genetic Cosmology

The raw data for generating empirical genetic cosmologies, and for testing
model genetic cosmologies, are collections of DNA sequences from a large
number of different species. From such compilations one proceeds as follows:

1. Order the species, evolutionarily, using classical phylogenetic methods
    or presently available phylogenetic methods [1–4].
2. From each species, select a DNA sequence that codes for a protein
    having essentially the same function in all species.

The result of this process is the construction of a series of homologically
ordered DNA sequences with an associated approximate evolutionary time
scale. Such a series can be used to suggest, in a rough manner, the functional
form of the fundamental tensor. Clearly, this will require much trial-and-error
model building. However, the difficulty with this ansatz is the paucity of DNA
sequences presently available. Detailed empirical research in genetic cosmology, therefore, must wait until such data become available.

4. Discussion

This work, formal though it may be, represents a significant simplification:
namely, attention is redirected away from the complex physicochemical
processes involved in evolution (as mediated through natural selection) to the
totally geometric concept of an evolutionary field generated by the curvature
of the informational space–time manifold. Such a treatment is crucial to any
construction of a theory in which the concept of biological symmetry is striped
of its imprecision. In view of our ability to represent the change in information
content of an evolving DNA molecule, we expect that this development will
facilitate future analyses of the nature of symmetry in biological evolution
formulated at the level of molecular genetics. Our intention in this chapter was
to show that these concepts are sufficient to account for evolution; the
demonstration of their necessity awaits further study.

Although quantitative research into the nature of the biologically correct
genetic cosmology is currently impracticable, there still exists a wealth of
classical evolutionary ideas which may lead to qualitative statements concerning the manifold structure. For example, biological evolution is generally

thought to be *divergent* [10]. At the molecular level this means that, starting from a single DNA molecule, two lines of evolutionary descent never terminate, at the same evolutionary time, in a DNA molecule which is the same for both lines of descent. In terms of the theory presented here, divergent evolution merely signifies that two evolutionary motions emanating from the same point never cross. In turn, this implies that the biologically meaningful subset of the informational space–time manifold cannot be compact.* However, in terms of the discussion of Chapter 8, we conclude that that subset of the manifold which has biological meaning cannot be closed.

Suppose, on the other hand, that these same two lines of descent terminate, at the identical evolutionary time, in DNA molecules which differ only in their codon order. Such an event represents the crossing of two evolutionary motions emanating from the same point. However, from an evolutionary point of view, divergence has not been violated. Thus if the manifold structure is such that evolution is divergent, then the above termination of the two lines of descent is not possible. This, of course, is a prediction that must be tested.

As another example, we cite the dependence of a gene's mutation rate upon the nature of the particular gene in question [11]. Such a dependence intimates that the model presented in Section 3.C is, perhaps, too simpleminded since it precludes any dependence of the fundamental tensor upon the codon coordinates.

These brief remarks should indicate the necessity of continued, qualitative research in genetic cosmology. In the end, however, the fundamental test of our viewpoint must await detailed empirical analyses.

References

1. M. O. Dayhoff, *Atlas of Protein Sequence and Structure* Natl. Biomedical Research Foundation, Washington, DC, 1972.
2. W. M. Fitch and E. Margoliash, *Science*, **155**, 279 (1967).
3. T. T. Wu, W. M. Fitch, and E. Margoliash, *Ann. Rev. Biochem.*, **43**, 539 (1974).
4. E. Margoliash, *Adv. Chem. Phys.*, **29**, 191 (1975).
5. F. Sanger, G. M. Air, B. G. Barrell, N. L. Brown, A. R. Coulson, J. C. Fiddes, C. A. Hutchinson III, P. M. Slocombe, and M. Smith, *Nature*, **265**, 687 (1977).
6. G. L. Findley and S. P. McGlynn, *Int. J. Quantum Chem., Quantum Biol. Symp.*, **8**, 455 (1981).
7. A. M. Findley, S. P. McGlynn, and G. L. Findley, *J. Biol. Phys.*, **13**, 87 (1985).

---

*A compact set of real numbers is both closed and bounded.

8. L. P. Eisenhart, *Riemannian Geometry*, Princeton University Press, Princeton, NJ, 1926.
9. G. L. Findley, "A Riemannian-Geometric Realization of Molecular Genetics," Ph.D. Thesis, Louisiana State University, Baton Rouge, 1978.
10. W. A. Beyer, M. L. Stein, T. F. Smith, and S. A. Ulam, *Math. Biosci.*, **19**, 9 (1974).
11. T. Dobzhansky, F. J. Ayala, G. L. Stebbins, and J. W. Valentine, *Evolution*, Freeman, San Francisco, CA, 1977.

Index

Abelian groups, 24–25
  multistructure objects, 78
Adaptive strategy theory of evolution, 129
Additive groups, 24–25
Adenine, genetic code evolution and, 33
Algebra:
  set theory and, 21–27
  vector and metric spaces, 5
Alternative coding:
  mitochondrial, 64–68
  symmetries of, 54–63
Alternative codings, in vitro, 66–67
Alternative or ambiguous codons, universality of GGC, 53–68
Ambiguous codon assignment:
  generalized genetic code symmetry, 48–49
  SGC symmetry, 39–41
Ambiguous or alternative codings, 37–38
Amino acids:
  ancestral nucleotides, 135–136
  degeneracy of, 30–32
  esterification of, 85–86
  eukaryotic cytochromes, 131–132
  genetic code and, 28–29
  mRNA translation, 84–85
  primordial, 33–35
  SGC symmetry, 38–41
  substitutions, 132–133
Amino acid starvation, alternative coding and, 66–67
Aminoacyl-tRNA, 84–85
  ribosomal binding sites, 86
Aminoacyl-tRNA synthetases, 85–86
Anagenesis, 129–130
Ancestral nucleotides, 135–136
Ansatz analysis, generalized genetic code, 70
Anticodon bases, genetic code and, 32
Antisymmetric tensors, 111

Associate tensors, 115
Automorphism, order isomorphisms and, 51

Bacterial cytochrome c, 131–132
Base pairing:
  complementarity of, 12
  DNA structure, 10–11
Biological context:
  GGC universality, 58–68, 68–69
  SGC symmetry and, 40–41
Biological evolutionary force, 100
  DNA base sequences, 139
  equations of motion, 125–126
  intrinsic and extrinsic, 126
Biological information transfer, geometric description, 5–6
Biological symmetry, theoretical considerations, 4
Biosynthesis of DNA, 12

Cardinality:
  of sets, 22
  vector space and, 94–95
Cartesian product, 23
  group structure and GGC symmetry, 44–45
Chargaff's equivalency rule, 11
Chemical mutagens, point mutations, 128
Chemical polarity, DNA structure, 11
Christoffel symbols:
  of the first kind, 118–119
  of the second kind, 119
  geodesics and, 141
Chromosomal mutations, 127
Cladogenesis, 129–130
Codon:
  amino acid-related, 30–31
  defined, 15
  nondegenerate, 43

151